HORSES OF DISTINCTION

Printed in the United States of America
Library of Congress Control Number: 2006922077
ISBN 0-9778947 – 0 – 3

HORSES OF DISTINCTION
STARS OF THE PLEASURE BREEDS

F. LYNGHAUG

HALLELUJAH
PUBLICATIONS
Downing, Wisconsin

Dedication

To the Maker of all horses, Who enjoys them tremendously and will come at the end of the ages, mounted on His horse. It was by His grace that this book was inspired and accomplished.

Table of Contents

"Ravenwood Keystone", 2005 PtHA World Champion Halter Gelding.
Owned by Cheyenne River Performance Horses.   Photo by Full Moon Photography.

Introduction

This book is about pleasure horse breeds that have certain definable and outstanding traits. It is for anyone interested in horses who would like to become better acquainted with them, their qualifications and their structure. It answers a need for specifics about horses, particularly when it comes to identifying breeds.

The best means of achieving this kind of accuracy was by calling on national North American registry information, which is the basis for the book. To verify breed characteristics, these organizations put forth a cooperative effort to compile and present their facts – an amazing feat!

The result is this fascinating and helpful reference for registering pure-breds. It has surprising insights on different breed features. It contains descriptions, histories and official standards of what each breed should be as presented by their registry.

It sets the record straight concerning horse breeds.

The Breed Standard

What are breed standards and why are they important?

Webster defines a standard as "an accepted measure of comparison for quantitative or qualitative value; criterion". Criterion is a standard on which a judgment is based. In other words, a standard is the epitome of a person or group's intelligent deduction of what the ultimate object is; what it contains, what it looks like, how it acts, the measure of its worth. A standard is the ideal of what something should be.

The beauty of a horse is directly related to how closely it resembles its breed standards. Observers who are impressed and marvel at a particular horse are, in essence, agreeing with the standards of that breed.

Moriesian: Tinus's Dream Weaver

er prestigious position and have much to offer their breed.

Purebreds are supposed to look and behave the way their standards describe. Those that deviate from that aren't considered in any way unique or better because they are different from others of the breed.

The difficulty is not in producing "unique" horses in a breed, but in producing many horses that closely resemble their breed standards. When this is achieved, it is a sign of a stabilized, recognizable breed and good professional breeding. Overall, the purest breeds typically display many horses with common traits. The gene base is proven and consistent. When a breed like this is mentioned, an immediate mental picture

Standards differ from breed to breed. What is an asset in one can be a serious fault in another. Therefore equine registries found it necessary to define their own individual breed standards. A knowledgeable breeder is familiar with his breed's particular standards and is constantly attempting to achieve them.

In dealing with the genetic nature of animals, perfection can never be attained, but the ideal standard is held up as the comparison of how close a horse can be to it. It is the ultimate goal. In the show ring, equines are judged by comparing them to their standards. Those that differ greatly from their breed standards, such as having an unacceptable color, are considerably distanced from what their breed is supposed to be. Those closest to it are considered to have better quality and hold more value. Individuals that adhere closely to their standards are typically regarded as more desirable for showing and breeding. They hold a high-

of what it looks like easily comes to mind. An observer is assured of what kind of equine is being presented.

The ironic twist is that an individual horse that has the traits of his breed to the highest degree is, in fact, unique because of the difficulty in producing this kind of quality. Gene combinations of sire and dam are always a gamble. Not until eleven months later will it be evident if the result mirrors what the pedigree says is there.

But the goal is to appreciate and value a horse whose characteristics correspond closely to its breed standards. Though there is no such thing as a perfect horse that fulfills exactly all its breed prerequisites, it's a wonderful challenge attempting to produce one.

Webster's first definition of a standard is a flag. And that is what the standards in this book are, a brilliant flag waving for all to see.

A Note About The Experts...

To best relay an honest breed presentation, it was important to use registry expertise because breed facts and standards usually originated with them. Registries have dated and pertinent facts not accessible to others. Their group efforts are consistent over a prolonged period of time to promote breed purity. They have accurate knowledge and a heart for the advancement of their breeds. They are the best recognizable authority and encyclopedia on equines.

Registries also provide the guidelines for accepted conformation. This is achieved through the many professionals, some of whom worked to establish their breed or participated in its development. Registries tap into this wealth of knowledge and experience to develop intelligent standards.

Even beyond this expertise, registries research facts, pool their sources, and agonize over correct standards to be as exact as possible with much thought and deliberation

American White Horse: TWHR's Tejas

– a truly mammoth project. Usually this was done without pay or reward of any kind, yet oftentimes at the risk of criticism and disregard of others. They did it because they loved their breed.

Consequently what they have to say about a breed is received as a serious definition. When entering a horse show or registering a horse, their qualifications are all that should matter.

Therefore, information was taken directly from them whenever possible, or through professional breeders closely tied to and assigned by those organizations when it wasn't possible. It was attempted not to change their wording or meanings in any way.

The participating breed organizations are the parent societies in North America. This insured consistent and correct information. It was through their extraordinary group cooperation and patience that this book was made possible. To them we are grateful.

Disclaimer

Some horse experts may contend that not all of the horse breeds presented in this book are true breeds. Genetically speaking, that may be true. Yet in general social equine circles, they have been accepted as breeds.

According to the United States Department of Agriculture, Handbook #394, "A breed of horse may be defined as a group of horses of common origin and possessing certain, well-defined, distinctive, uniformly transmitted characteristics that are not common to other horses…" (Thanks International Morab Breeders' Association!)

The purpose of this book is to honestly depict the traits usually common to a particular kind of equine. For that reason, it was attempted to give equal attention to all the breeds as represented by the breed organizations and to take each one seriously. I didn't presume to decide which breeds were "authentic" and which were not.

It may also seem that there are several representations of the same breed. If attention is focused on the breed standards, it will be evident that there are differences.

My hope was to define the various breeds or types as accurately as possible because they are so wonderfully diverse and beautiful in their own unique ways.

F. Lynghaug

Breeds In This Book

International Colored Appaloosa: Oak's Lyon Heart

    Horses of Distinction are notable North American riding horses possessing outstanding traits.

    In general, they are light pleasure riding horses that are spectacular and unique. They display special features or discriminating characteristics particular only to them. They may have a distinctive coat or coat color. Some have certain head properties or a tail carriage that is exclusively theirs. Others have a flashy presence, displaying their own kind of animated glitz and glamour.

    They project a special flare and pizzazz all their own. They belong in the spotlight. They make that "Hey, look at me!" impression. Some sparkle in the show ring and some prance in parades. But all have noticeably visual attributes that cause them to stand out from the other light breeds. Their presence draws considerable attention and sets them apart in a class of their own.

    They are horses of exceptional distinction.

Akhal-Teke

The magnificent Akhal-Teke comes to us intact in all his glory after 10,000 generations of breeding on the Central Asian steppes. Since the time of earliest horse domestication, this "Golden Horse" of Central Asia has been prized before all other breeds. Bejeweled and decorated in gold and silk, he was ridden into battle by nomads and emperors and buried with honor in the tombs of kings, shamans, and warriors. In Chinese legends he was known as "the Heavenly Horse"; the Han Chinese felt it well worth 80,000 soldiers to obtain only 20 Akhal-Tekes, such was their reputation.

central figure in every culture into whose hands it came.

There were excellent reasons for the horse's many stewards to carefully preserve him. There were no other horses that could run faster or go greater distances on such little or total lack of food and water for days on end. No other horse was more intelligent or more devoted to its rider. No other horse had the glittering colors that helped earn it the name "Heavenly Horse".

Characteristics:

The Akhal-Teke (pronounced "ah call tek ee") is a true desert bred horse with a light, elegant build and distinctive conformation. It is long, lean, and typically narrow through the chest, making for an extremely comfortable ride. A characteristic feature is the sparse, short mane and forelock and absence of feather on the legs.

And yes, Akhal-Tekes really do glow like metal! They have a unique hair structure which refracts light, displaying colors from blazing palomino to electric black. Glittering gold is their prevailing color.

The Akhal-Teke retains every quality of endurance, speed, economy, intelligence and beauty that was so prized throughout the centuries by so many different societies.

The breed originated in Southern Turkmenistan as the chief mounts of Turkoman warriors. At the Akhal oasis a distance from the main trade routes, the Teke tribe first bred them. Living in the desert, Akhal-Teke horses developed endless stamina and the ability to withstand great extremes of temperature and deprivation.

Five successive empires – the Scythians, Parthians, Ywati, Huns and Turkmen – invaded the area, laying waste to everything before them, yet they carefully preserved these magnificent horses by keeping and training them with the utmost care. In time the beautiful Akhal-Teke became the

In a fast paced world where a horse is often little more than a living vehicle or a piece of sports equipment, the Akhal-Teke offers a refreshing change. For three millennia, the Turkmen bred their horses to be more than transportation, but a part of their families. Horses lived tied to the tent and were ridden often in search of a living. Their reactions were used for "testing" would-be suitors for daughters of the family. They were fed by hand by the entire family and were included in the count of the family's wealth.

Thus the Akhal-Teke has become sensitive, sensible and intelligent. It learns quickly and with enthusiasm. It is very much a horse that seeks to bond with a person "of his own", forming a lasting partnership. While most Tekes are friendly to everyone, they tend to respond best to their regular rider, often performing better for "their" person than for a casual rider with better skills. One may start out owning an Akhal-Teke, but in the end, the Akhal-Teke will own the person! Memories of a first experience with an Akhal-Teke remain for a lifetime.

Today the role of a Teke has changed from war horse to sport horse and his unique character suits him well to a variety of disciplines. Akhal-Tekes have won Olympic gold and silver medals in dressage and have been jumping champions throughout Europe.

In Germany, Akhal-Tekes compete in reining and Follow-the-Hounds. In the United States, they are used in eventing, jumping and endurance. The Akhal-Teke's exceptionally efficient heart and lungs help with the "equine radiator" effect; they show some of the quickest cardiac recovery scores of any breed at vet checks on endurance rides.

In Canada, Akhal-Tekes are making their mark in the sport of Competitive Trail, where their legendary fearlessness and trust of their rider makes them the perfect team player.

All over the world, the Akhal-Teke shows its ancient prepotency in producing purebred and part bred foals of exceptional ability, strength, soundness and beauty.

The United States is very fortunate to have an ever increasing number of breeders dedicated to preserving the very best of this marvelous and valuable horse. Nowadays, it is possible to see examples of the Teke all across the country in competitions and on breeding farms. Best of all, with a wide choice of superior stallions available at reasonable fees, Teke blood is now available to everyone.

The Akhal-Teke Association of America oversees two registries: The Akhal-Teke Registry of America and the Akhal-Teke Sporthorse Registry of America for crosses of one half or more of Akhal-Teke blood. The Sporthorse registry is not for the creation of a new breed, but to record sporthorses whose breeders have utilized the Akhal-Teke gene pool in their bloodlines. Akhal-Tekes are registered on the basis of parentage only.

BREED BASICS

A true desert horse, the Akhal-Teke is aristocratic, graceful and elegant with an underlying strength and intelligence that sets him apart from other breeds. Being of such an ancient lineage and due to its geographical isolation (precluding any infusions of pony or draft blood), the Akhal-Teke retains the features of its earliest ancestors.

If the Akhal-Teke was bred over the centuries for any particular purpose, it was to get riders between waterholes quickly and comfortably. Typically these waterholes were 80 miles or more from each other. This use resulted in a comfortable, swift, sleek horse quite different from more fashionable breeds that are common today.

The Akhal-Teke's overall impression is one of an elegant, exotic animal exuding grace, power and athleticism. The comparison in appearance to a cheetah or fine greyhound is not inaccurate. While degrees in type are allowed and even encouraged, all examples of the breed should carry the distinct characteristics that differentiate the Akhal-Teke from other horses.

In judging or grading Tekes, basic soundness is of primary importance, followed by the presence of type. Severe faults in conformation shall be penalized severely even in the presence of outstanding type. Brilliance in type, conformation and/or movement, though possibly accompanied by minor faults or shortcomings, shall be recognized and rewarded highly.

General Conformation:

The Akhal-Teke is meant to be a medium sized horse, ranging in size from 14.3 to 16 hands. Extremes in either direction are not desirable. In general, the Akhal-Teke gives the impression of length, without showing weakness or frailty. It should be longer than it is tall, with a rectangular silhouette.

The back is long, but strong, with a level topline. Withers are high and prominent and attached to a well set-in shoulder. Shoulders should be long, nicely sloped and extremely free-moving. The wither height combined with the relative narrowness of the chest at the shoulders makes for an enormous range of motion of the shoulder, giving the horse even greater scope and power.

Although the chest is narrow when viewed from the front, the heart girth is deep. The barrel widens smoothly out to the hips with little curvature of the ribs. The hip angle is wide and gives the appearance of strength. Tailset is low.

Faults: Extreme heaviness or reediness. Excessively long back, especially when coupled with a weak loin connection. Extreme downhill conformation. Mono or cryptorchidism. Thick, coarse or overly muscular appearance. A square outline, the horse being taller than it is long.

Head and Neck: The Akhal-Teke's head is long and narrow with most of the length being from the eyes to the muzzle. It has a straight or slightly dished profile. Overall, the head is dry (finely made and without excess flesh), with large nostrils and thin lips. Eyes are large, expressive and often hooded or "oriental" in appearance. Ears are long, slim, mobile, set forward and very alert. The throatlatch is refined, the poll is flexible and the long, slim neck is set high out of the shoulder.

Faults: Severe overshot or undershot jaw, common or coarse head, thick throatlatch, thick neck, low neck set.

Legs/Feet: The Akhal-Teke is a true desert horse, and as such, should possess extreme stamina and hardiness. The presence of adequately dense bone is one such indicator of these traits. Akhal-Tekes have short cannon bones and low-set hocks, while the forearm and gaskins are long and smoothly muscled. Legs are dry with tendons well-defined. Joints are large. Knees should be flat. Pasterns should be long and display an identical angle to the hoof and shoulder. Hooves are small, round and extremely hard.

Faults: Any and all limb formations that could contribute to future unsoundness, including but not limited to: bench knees, calf knees, offset cannon bones, sickle hocks, wide at the hocks, lack of bone, small joints, pigeon-toes or toed-in stance and dished hoofs. Horses shall be penalized according to the severity of the fault.

Coat: The skin of the Akhal-Teke is very thin, with their coat and hair being quite fine and silky. In summer the skin around the eyes and nostrils may be bare. Tekes often have a sparse mane and tail, little or no forelock and the absence of feathering on the fetlocks.

Movement: The Akhal-Teke is today what he always was – a horse bred to go very long distances. Consequently he does not have the spectacular movement that seems to be currently fashionable in many breeds. Yet his movement is one of his most desirable breed characteristics. It is unique and highly distinctive. His gaits are long and sweeping without the unnecessary elevation of knees and hocks. While not truly gaited, he has long, energetic strides that float above the ground at the trot and sweep flat to the ground at the canter, making this horse an extremely comfortable ride.

His action is quite forward in all three gaits. His low, fluid trot and sweeping canter have little swing or knee action. Instead, all movement is free flowing and elastic, as if he is sliding or skimming just above the earth. Overall, the effect is that of a fine greyhound.

Because of the Teke's characteristically high head carriage, he is naturally more balanced toward his hindquarters. Impulsion is not something you have to beg for in a Teke! He has magnificent action, gliding with soft, expansive strides.

The Akhal-Teke has a different style of jumping than other breeds of horses. His style is more like that of a deer with head and knees up rather than out. Thus he puts his center of balance over his heart rather than his withers or shoulders. He is a naturally careful jumper; he likes to leave a good deal of daylight between himself and any rail. He is very talented and agile, jumping with fearlessness and endurance unmatched in any other breed of horse.

The Akhal-Teke gives the impression of lithe athleticism without excessive musculature.

Faults: Winging, paddling, excessive knee action, heavy or ponderous gaits, lack of forward drive.

Character: The Akhal-Teke is a very intelligent, hot-blooded horse who develops a dog-like devotion to his rider. He has a quiet temperament, but is easily aroused. He is bold, tenacious, tough, resilient, and genetically conditioned to extremes of heat and cold. Tekes can withstand great privation and come back into form readily when conditions improve.

COLORS:

Any color is acceptable in the breed, as is any combination of white markings. Although all colors occur, golden buckskins and duns are the most prominent.

Typically there is a marvelous metallic glow to the coat. This characteristic gleam is a desirable feature. It is seen most noticeably on the buckskins, palominos and duns. It is caused by the structure of the hair; its opaque core is greatly reduced in size and may in areas be altogether absent. The transparent part of the hair (the medulla) takes up this space, and acts like a light-pipe, bending light through one side of the hair and refracting it out the other side, often with a golden cast.

Few breeds of horse can claim the diversity of coloration to be found in the Akhal-Teke. While some breeders prefer certain colors over others, there are no disallowed colors or markings in the Akhal-Teke breed.

Variations:

Dominant black, also called "electric black" and "raven black", is quite common in the breed. Combined with the famous Akhal-Teke glow, these horses literally glitter with a blue or purple sheen. This color is so special that it has its own name, "voronaya", in Russian. This is the color of five time Olympic medallist (2 gold, 2 silver, 1 bronze) "Absent", a purebred Akhal-Teke stallion.

Light palomino is often called "isabella" in much of the United States, although in Europe and Russia "isabella" refers to cremello and perlino. A famous light palomino Akhal-Teke is "Kambar", the world record holder for racing distances of 4000-8000 meters. He generated interest and the start of Teke-love for many an American.

Mahogany bay has the "sooty" factor, which gives a black tip to the hairs, producing many lovely variations in horse color. "Astrachan" is a gorgeous mahogany bay and the Number One rated Elite Akhal-Teke stallion in North America.

Dark golden dun along with golden bay, golden buckskin and golden dun, are some of the archetypical Akhal-Teke colors. In the sunlight, the coat glitters with gold in a way that a camera simply cannot capture. This color (actually a form of olive grulla) is so dark that it is often mistaken for a non-red bay, but a stripe down the spine and zebra striping on the legs show that it is a dun.

Bay can be quite spectacular with coats that glitter with reds and golds.

Dunskin in the Akhal-Teke can combine the dun and cream factors in some striking ways. Usually these horses have black manes and tails, but they can also have bicolored manes and tails.

Liver chestnut, while not as common in the Teke as it is in other breeds like the Morgan, is still found quite often. In fact, this is the color of the line founding stallion, "828 Fakirpelvan", sire of the famous European jumper, "Penteli".

Palomino has some truly stunning examples in the Tekes, usually with plenty of "chrome" (white markings.)

Chestnut tends to be more golden than red, but there are a fair share of lovely red chestnuts.

Cremello, (or "isabella" in Europe) and perlino are found quite commonly in the Akhal-Teke. The glow to the coat of these blue-eyed wonders is so strong that it is visible even in a darkened barn.

Claybank or red dun is a rarity among Tekes.

Cream grulla is a combination of grulla or dark golden dun and perlino. In this color, the eyes may be blue, gray, or hazel.

Grays are actually fairly common in Tekes. They are often beautifully dappled and many turn completely white.

Grulla, among Tekes, are of the "olive" variety and usually called dark golden dun. Grulla can be distinguished from gray in that the head is dark and the color does not change with age.

Golden, called "bulanaya" in Russian, is the archetypical color of the Akhal-Teke horse. These horses are sometimes purely buckskin (without a spinal stripe) or dunskin (with a spinal stripe a few shades lighter than the mane and tail but darker than the coat and plainly visible). Because of the unique structure of the hairs, these horses may be quite dark in color and may even be confused with bay; however, when bred together they can produce perlinos and cremellos.

Perlino differs from cremello in that there is some reddish or brownish color to the tail and often on the hocks, knees and legs. As with the cremello, the eyes are blue.

Akhal-Tekes typically have white markings, and some sport a great deal of chrome. Sabino pinto markings are not at all uncommon.

Rabicano is also seen. This is a form of roan which shows white only where the skin is very thin or particularly stretchy: on the flanks, over the ribs, or on the throatlatch. When it's on the ribs it's usually in the form of vertical stripes, but generally it's seen only on the flanks or sometimes only on the little flap of skin just over the stifle. Older stud books list roan as an Akhal-Teke color, although this line seems to have died out.

American Bashkir Curly

Owners of Bashkir Curly horses get used to comments like, "How long does it take to curl that hair?" "Did you give your horse a perm?" or the best one yet, made by an old cowboy who accused an owner, "What the ___ did you do? Cross a horse with a sheep?" They get used to it because the American Bashkir Curly always draws a crowd wherever it goes with its body covered in curls. People want to touch the unusual coat, which feels like lambs wool. They sink their fingers into it instead of petting like they would a regular horse. Some put their faces up to the curly coated neck to feel it better. One little boy played with a curl, making a "spruuung" noise as he pulled and released it repeatedly like it was a spring. It delighted him for quite some time while the Bashkir Curly patiently stood there.

Those who know the breed appreciate it because it's different. But there is more to the Bashkir Curly than the curly coat. Its versatility, soundness and docile temperament endear it to owners. The curls are unique and draw attention, yet it's their personality and diversity that causes continuing popularity.

HISTORY:

The foundation for today's knowledge of the Curly horse stems from the work of one family in Nevada, the Damele's. In 1879 Giovanni John Damele emigrated from Italy to the United States and in 1898 he purchased the Three Bars Ranch in Nevada. His grandson, Benny Damele, noted it was probably about this time that they first spotted curly coated horses running in the wild herds near their ranch. However, it wasn't until 1931 that two of the Damele boys caught their first curly coated horse, a sorrel, in the Roberts mountain range. They broke him to ride and sold him.

Then in 1932, the Great Basin of Nevada was hit by an extremely harsh winter. Most of the domestic stock and wild horses either starved or froze to death. Among the few survivors were the "Curlys". It became apparent to the Damele's that these horses had special features which enabled them to survive. They were also gentle, easy to train and could work cattle. At this point the Damele's made an assertive effort to capture and breed the Curly horses.

In 1942, Peter Damele moved his family to the Dry Creek Ranch in Austin, Nevada, just south of Three Bars. Dry Creek is the ranch now associated with the first domestic breeding of Curly horses. It's horses carried the 3D brand. Descendents of these horses have come to be known as the "Damele line".

In 1971, Benny Damele (son of Peter Damele) and a fellow Curly enthusiast, Sunny Martin, organized the American Bashkir Curly Registry in hopes of preserving the remarkable horses. Because of the small number of horses that they had to work with, inbreeding was a major concern. In an effort to introduce new blood, they chose to cross their Curlys with the following four breeds:

ARABIANS: A breed that shared the same short back as the Curly, indicative of a five lumbar vertebrae spine, and who were also known for their endurance.

MORGANS: Horses whose conformation was very similar to that of the Curlys.

APPALOOSAS: Native American horses that had been bred for endurance and who shared the unusual Curly trait of shedding mane and tail hair.

MISSOURI FOX TROTTERS: A breed that Sunny felt shared the smooth gait of the Curly. Today some people feel that the only "gait" exhibited by the original Curly was actually a "running walk" or "Indian shuffle"….not the full gait of the Missouri Fox Trotter. This issue has never been proven one way or the other, but this cross produced some beautiful, gaited Curly horses. Many breeders today breed specifically for the "Gaited Curly".

Some of the most notable "outside" studs used were: "Nevada Red" (Arabian); "Ruby Red King" (Morgan); "Chocolate Chip D" (Appaloosa); and "Walkers Prince T" (Missouri Fox Trotter).

Research continued and it was discovered that the Curly horses were not limited to Nevada. The Crow and Lakota Sioux had also been capturing and riding Curly horses in the plains of Montana and South Dakota. Proof could be found on American Native drawings. (Most Native American Indian tribes had kept records of important events that occurred during each year with drawings, usually on animal hides. White men called these records "winter counts" because the Indian year started in the spring and ended at the conclusion of winter.) The Sioux winter count of 1801 drawn by the Lakota Sioux Indian, High Dog, depicts the Sioux stealing Curly horses from the Crow Indians. There were several winter counts found containing the same information.

In 1881, Chief Red Horse, a Minneconjou Sioux, was persuaded to make some drawings of the Battle of the Little Bighorn. One of those pictures depicted an Indian riding a curly coated horse.

Descendants of these horses are considered the Native American line.

It was originally believed that the Russian Bashkir horse was curly coated and was probably an ancestor of the American Curly. ("Bashkir" refers to a region in Russia.) Hence, the name AMERICAN BASHKIR CURLY. After the fact, it was discovered that it was not the Bashkir horse that had a curly coat. It was actually the Lokai, a breed of horse from the Lenisske Region of Tafjik, U.S.S. R. (Russia). Members of the American Bashkir Curly Registry considered dropping the word "Bashkir" from the name but after lengthy discussions, for the sake of public recognition, they left the name as it was. Many breeders, however, have chosen to refer to their horses as simply "American Curlys".

Just how did these unique horses find their way to America? The following three theories were explored by Shan Thomas in his book "The Curly Horse in America – Myth and Mystery":

1. Were the horses imported and left behind by the Russians who settled the fur hunting colonies along the Pacific Northwest? No evidence can be found to support this.

2. Were they imported into central Nevada by a rancher named Tom Dixon in the late 1800s? Research produced information that Tom Dixon purchased a Curly stud and two Curly mares in northern India and imported them to Nevada. He settled in Eureka, Nevada and used feral herds in his breeding program. That would certainly account for the Curly horses' presence in Nevada. However, that does not account for their presence in the Dakotas in the early 1800s.

3. Did the horses of ancient origin come across the Bering Straits land bridge thousand of years before the Spanish introduction of horses?

There are no fossils to substantiate that theory.

Shan Thomas of the C.S. Fund Conservancy completed his research without proving any of the theories. Genetic research is currently being conducted by Dr. Cecelia Penedo of the University of California Davis and Dr. Gus Cothran of the University of Kentucky. Hopefully, some of the mystery will be solved in the near future.

In the late 1990s, the number of registered Curlys was approaching 3000. There was talk of closing the registry books as many feared that continued crossbreeding would dilute the gene pool to the point where the Curly's unique characteristics would be lost. Others opposed the idea as they felt the gene pool was still too small. Many breeders were still crossbreeding in an attempt to create the "perfect horse". As one old cowboy said, "They ain't improvin' the Curly horse none…they're improvin' the OTHER breed!"

The Registry consulted with Dr. Ann Bowling, genetic researcher at the University of California, Davis. She stated that, in her opinion, there were a sufficient number of registered Curlys to support closing the books. After many long discussions, the American Bashkir Curly Registry closed its book in the year 2000. As a result of that decision, the ABCR is now a blood registry as opposed to a coat registry.

Knowing that there were many fine curly coated horses that would no longer be registerable, a second registry, "The International Curly Horse Organization" was founded. ICHO continues to register all horses with a curly coat, regardless of pedigree, which allows the Bureau of Land Management horses (feral range horses) and crossbreds to be tracked. This provides a valuable service to all Curly breeders.

Characteristics:

Other than the obvious curly coat, the Bashkir Curly is valued for its gentle disposition, high intelligence, strong bone, soundness, versatility of performance, strength and endurance. As a result of the years of cross-breeding, the Curly can be found in every size and color: mini to draft…solid color, pinto or appaloosa.

The Curly is famous for his ability to endure hardship and would definitely be classified as an easy keeper. However, it still needs all the care and comforts that would be offered to any other horse. Curlys do need to be wormed, vaccinated and have their feet cared for.

As for that curly coat….The coat is not typical horsehair. Testing has proven it to be more closely related to mohair; it can be spun and woven into garments. The hair is also considered (but not proven) to be hypoallergenic. People who are allergic to straight haired breeds usually will not have a reaction to Curlys.

The winter coat expresses itself in a variety of patterns commonly described as marcel wave, crushed velvet, curl or micro curl. The summer coat also offers varieties ranging from smooth to wavy. Some horses shed their manes and tail hair every year, only to grow them back in the winter.

Not all Curlys have curls over their entire body. A lot depends on the temperature – the colder the weather the longer the hair and the heavier the curl. Genetics no doubt plays a part in this, too.

The official breed identity standards are as follows:

BODY CONFORMATION: A medium size head with well-defined jaw and throatlatch. Wide set eyes with eyelashes that curl up. Ears that are short to medium in length with curls inside; the ear hairs do not totally shed out in the summer. A medium length neck; deep at the base of the neck where it joins the base of the shoulder. Noticeably short back; deep through the girth. An appearance of a long underline and belly cut high in the flank. Heavy boned legs and short cannon bones compared to the forearms. Forequarters and hindquarters should be supple, yet well muscled. Withers are medium. The croup should be flat or with a shallow slope to the base of the tail. Curlys should travel (move) easily and smoothly.

TYPES OF CURLS ON BODY:
   Fine, soft hair.
   Ringlet (can be several inches long).
   Marcel Wave (deep soft wave in the body coat)
   Crushed Velvet (soft dense pile of curls in body coat)
TYPES OF CURLS ON MANE:
   Extra fine hair.
   Kinky (is preferable)
   Wavy
 TYPES OF CURLS ON TAIL:
   Ringlets
   Wavy

TYPES OF CURLS ON FETLOCKS:
>   Curly (is preferable)
>   Wavy

MANE: May shed in the summer.
>   Split mane (hangs on both sides of neck).

TAIL HAIR: May shed at the head of the tail in the summer.
>   Some Curlys may shed additional tail hair.

FETLOCK HAIR: Should shed in summer but still retain some
>   long hairs.

HOOVES: Almost perfectly round in shape. Very hard and dense.
>   Proportioned to the size of the horse.

SIZE AND WEIGHT:
>   Average – 14 to 16 hands.
>   Average weight – 800 to 1000 pounds.
>   Some Curlys may weigh up to 1200 pounds.

UNIQUE CHARACTERISTICS:
>   Even tempered disposition.
>   Quick pulse and respiration recovery.
>   Very dense bone in the legs.
>   Curly coat can be hypoallergenic. The hairs are round instead of flat.

They appear barbed or feathered underneath a microscope and can be spun and woven.

Can shed some of the mane and tail hair in the summer, but they grow back in the fall with the winter coat.

Noticeably short, strong back indicating a five lumbar vertebrae.

Due to the years of outcrossing, Curlys can be found with differing conformation, sizes and weights. This influence will probably continue for the next five to ten generations of Curly-to-Curly breeding. However, these horses will be registered as long as they meet ABCR criteria.

Curly horses have excelled in all disciplines – working cattle, trail riding, endurance, and showing in both English and western. Therapeutic riding schools appreciate their gentleness and have been searching out Curly horses for their programs. Curlys are always a favorite in parades and at equine fairs. Every year the "Classic Curly Riders" can be seen in the Rose Parade in Pasadena.

Knowledge of the Curlys now extends beyond the American and Canadian borders with several having been exported to European countries. Their popularity in Europe is growing by leaps and bounds.

If they sound too unique to be true, the only way to appreciate Curlys is to see one up close and personal.

American Paint Horse

An ideal horse's best qualities should be intelligence, willing disposition, sound conformation, versatility, athletic ability and strength. Add to this some beautiful colors in a seemingly infinite number of distinctive coat patterns and it produces one of the most popular breeds in the world, the American Paint Horse.

These desirable qualities were first and foremost on the minds of the breed's founders. As a result of the efforts of tens of thousands of people who have been involved with Paints over the past 44 years, the American Paint Horse has become one of the greatest horses in the world. Its popularity has grown to include Paint owners in about 40 nations and territories as well as the United States. Its breed registry has become one of the fastest-growing, issuing pedigree certificates at a rate of about 50,000 every year.

HISTORY:

In 1519 the Spanish explorer Hernando Cortes sailed to the North American continent to find his fame and fortune. Along with his entourage of conquistadors, he brought horses to help his men travel across a new world in search of riches. These excursions left behind a profound legacy – the bloodstock that would provide the foundation for a variety of unique, distinct, American bred horses. According to the Spanish historian Diaz del Castillo who traveled with the expedition, one of the 16 war horses that carried Cortes and his men was a sorrel and white horse with spots on its belly. It is thought that this spotted horse laid the foundation for what is today the American Paint Horse breed.

By the early 1800s, the Plains were generously populated by free-ranging horse herds which included similarly spotted horses. Because of their color and performance, the flashy spotted horses soon became a favorite mount of Native Americans. The Comanche, considered by many authorities to be the finest horsemen on the Plains, favored loud colored horses and had many among their immense herds. Evidence of this favoritism is exhibited by drawings of spotted horses found on the painted buffalo robes that served as records for the Comanche.

Throughout the 1800s and late into the 1900s, spotted horses were called by a variety of names: pinto, paint, skewbald or piebald. In the mid 1900s, groups began to organize in an effort to preserve these peculiar animals.

Rebecca Tyler Lockhart was one of the admirers of beautifully colored horses and horses with sound Western stock horse conformation and she sought the best of both worlds. Overcoming almost insurmountable odds and nearly unanimous opposition among breeders, Lockhart forged ahead with her plans for a new breed standard.

In 1962, she called a meeting with a special group of spotted horse enthusiasts in Gainesville, Texas, and they organized the American Paint Stock Horse Association (APSHA). It was dedicated to preserving horses of color with stock-type conformation. Following a merger with the American Paint Quarter Horse Association in 1965, it became the American Paint Horse Association (APHA).

This group thought the varied, distinct coat patterns of the American Paint were appealing. Being knowledgeable devotees of Western stock-type horses, however, they insisted that stock-type conformation had to be a major criterion for establishing a registry. The Western stock horse was appreciated for its talents in working with livestock as well as its strength, intelligence, speed and agility. It was typically a well-balanced and powerfully built animal.

Lockhart remembered seeing a continuing trend of discrimination against horses such as Paints with distinctive color patterns. "People looked down on Paints," she remarked. "They just didn't understand. They thought (Paints) were inferior. If a breeder (of solid-colored horses) had a Paint foal, he was ashamed. He thought it was a bad reflection on his herd, that it was connected with inferior blood. He was afraid someone would think there was something wrong with his breeding program. So a lot of ranchers would get rid of them." She recalled that Paints would often be relegated to the back pastures, or in some cases, suffered a worse fate.

"There were many good Paints that were destroyed back then," agreed Junior Robertson of Waurika, Oklahoma. Robertson was one of the few ranchers in the country who actually admired and owned Paints at the time. One of the most notable horses Robertson owned even before there was an association for Paints was "Wahoo King." This colorful sorrel Paint set the standard for top competition horses in the 1960s and was well known as a legendary roping horse.

Today the American Paint Horse breed standard has become widely respected and admired. The breed is now recognized not only for its beautiful coat color patterns, but also for its sound conformation and musculature as well. It is also highly regarded for its superior intelligence and willing disposition, making it an easily trainable horse.

While the colorful coat pattern is essential to their identity, American Paint Horses have strict bloodline requirements and a distinctive stockhorse body type. To be eligible for the Regular Registry, horses must come from stock registered with one of three recognized organizations: the APHA, the American Quarter Horse Association or the Jockey Club (U.S. Thoroughbred Registry). At least one parent must be a registered American Paint Horse.

The APHA also maintains minimum color requirements for registration in the Regular Registry. Stallions, mares and geldings that meet the bloodline requirements and have a "qualifying area" of solid white hair (with underlying unpigmented skin) on a non-white horse, or dark hair (with underlying dark skin) on a white horse with unpigmented skin, are placed in the Regular Registry. Spots must be on the horse at the time of birth.

Those horses that do not have sufficient qualifying spotted areas for the Regular Registry, but meet the bloodline requirements, can be considered for the solid colored Paint-Bred Registry.

Spotting patterns of Paint Horses have an added fascination over that of most other breeds. They not only have the colors of other breeds, but superimposed over these colors are a variety of beautifully individualistic spots.

Each Paint Horse has a unique combination of white patches and any background color of the equine spectrum: black, bay, brown, chestnut, dun, grullo, sorrel, palomino, buckskin, gray, cream or roan. Markings can be any shape or size and located virtually anywhere on the Paint's body.

Although Paint Horses come in a variety of colors with different markings, there are only three specific coat patterns recognized in the Regular Registry: tobiano, overo and tovero.

The tobiano pattern is distinguished by oval or round spots that extend down the neck and chest of the horse. White often crosses the horse's back between its withers and tail. The tobiano's head markings may be completely solid, or have a blaze, strip, star or snip. Generally the legs are white, or at least they are white below the knees and hocks. The tail is often two colors.

The overo pattern can be mainly dark or white. Typically, the white will not cross the back of the horse between its withers and tail. Overos often have bold white head markings, such as a bald face. Also, an overo's body markings will be irregular and scattered and one or all four legs will be dark. The horse's tail is usually one color.

Overo patterns can also be broken down into three different spotting patterns. "Frame overo" patterns usually have large amounts of white on the head and white spots arranged horizontally on its sides and neck. The feet and legs of this horse are usually dark, although they can have white feet or socks just as the non-spotted horse can. Frequently, frame overos have one or two blue eyes.

"Sabino" patterns have white on the legs and head and usually creep up on the body in the form of belly spots. They are usually flecked and roaned, although some are crisply spotted. Sometimes sabino horses have blue eyes.

The last pattern is much rarer than the others. It is the "splashed white" pattern. These have white legs and bellies, as well as a great deal of white on the head. The edges of the white are quite crisp. Many splashed white horses have blue eyes.

Because not all coat patterns fit precisely in the tobiano or overo categories, the APHA expanded its classifications to include the tovero pattern, which describes horses that have characteristics of both the tobiano and overo patterns.

These patterns—tobiano, overo (frame overo, sabino, and splashed white), and tovero are the color patterns that distinguish Paints from other horses.

Their colors, markings and patterns, combined with stock-type conformation, athletic ability and agreeable disposition, make the American Paint Horse an investment in quality. Owning an American Paint Horse is an enjoyable, positive experience.

The breed is experiencing an increase in popularity. There are now more than 1,100 approved APHA shows every year around the world. World Championship Paint Horse Shows can run up to 14 days with thousands of horses participating.

Paints have a high profile by proving at several venues around the nation that when it comes to intelligence, versatility and athletic ability, the breed is at the top of the class. Paints continue to stand out in reining with impressive victories at competitions such as the United States Equestrian Team Opening Reining Championship, the National Reining Horse Association Futurity and the prestigious United States Team Roping Championships. The Pacific Coast Cutting Horse Association Futurity and the National Cutting Horse Association Futurity are other competitions where Paints have proven themselves.

Since 1966 when APHA officially recognized the sport of racing, the Paint racing industry has made major strides forward. In that inaugural year, 17 starters ran for $1,290 in just two states – Texas and Oklahoma. But in 2005, more than 600 starters competed in nearly 800 APHA recognized races for purses totaling more than $4.5 million. A total of 18 states feature Paint racing, including Arizona, California, Colorado, Idaho, Kansas, Louisiana, Michigan, Montana, Nevada, New Mexico, North Dakota, Oklahoma, Oregon, South Dakota, Texas, Utah, Washington and Wyoming. Paint racing is also offered in Canada.

The American Paint Horse has established itself as an accomplished athlete in and out of the show arena. Whether racing, trail riding, or performing before thousands at worldwide competitions, the American Paint Horse's combination of beauty and performance are unmatched.

American White Horse and American Creme Horse

American White Horse

American Creme Horse

There is nothing so eye catching as white horses. They are like a neon sign flashing for all to see. White horses seem to naturally have their own spotlight constantly shining on them. They grab attention wherever they go. Audiences focus automatically on white horses before anything else; there is an absolute fascination with them.

Other horses seem to blend into the background compared to white horses. The whiteness of their coats glows like brightness in the dark. It doesn't seem to matter what color tack white horses wear in a show or parade, they just catch the spectator's eye, irresistibly drawing them in.

The Emperor of Japan rode a white horse. Noted country music singer, Tex Ritter, rode a white horse, "White Flash", when he initiated his career by first appearing in western movies. The Lone Ranger rode his white horse, "Silver", in a TV series. White horses are stately and picturesque. They are often mistaken for statues in photos.

"They make a definite statement about beauty with just their presence," says long time White Horse breeder, Jo Ann Anderson, "and a white horse in motion is mesmerizing!"

Cream colored horses experience the same fame as they are often mistaken for whites. Unless a Creme Horse stands alongside a White Horse, the difference can't be noticed.

In 1906 a white horse known as "Old King" was foaled. He was owned by Professor Newall of Illinois and was used for breeding circus horses. He was a snow white, pink-skinned, brown-eyed stallion of unknown breeding. He matured at 15.2 hands and weighed approximately 1200 pounds. He had a thick neck, stocky build, strong legs, a long, heavy and wavy mane and tail, and great intelligence with a good disposition.

In 1917 he was bought by Caleb Thompson and his twin brother Hudson, successful cattle ranchers. Crossed with their herd of predominately Morgan mares of various dark colors, King produced white foals much like himself. In time the ranch established the American White Horse breed and a white horse breeding program. Caleb and his wife, Ruth, bought out Hudson's share of the ranch and began showing their horses at local events. They originally named their new horse "American Albino" and organized a breed registry in 1936.

In 1937 they incorporated the American Albino Horse Club (AAHC), a registry and stud book for white horses. It eventually included a subsidiary, National Recording Club, in which all non-white foals were recorded, including cream colored horses as non-white or off-colored. (However the American Creme wasn't officially recognized until about 1980.) The registry eventually moved away from the use of the word "albino" in succeeding title changes since, genetically, the American White Horse is not a true albino and neither is its close cousin, the American Creme Horse.

In 1938, the Thompson's moved to a ranch of nearly 2,400 acres in Naper, Nebraska and named it El Rancho del Caballo Blanco. However, the local residents called it the White Horse Ranch. The name stuck and a legend began. The Thompson's began a training and riding school for children and started touring with their White Horse Troupe, doing performances that included the children along with the horses from the school. In the 1940s and 1950s the White Horse Troupe became internationally famous and toured all over the United States and Canada. The popularity of the White Horse Ranch also attracted many visitors.

In 1945 Life Magazine visited the ranch and ran an article on it in their August issue. The White Horse Ranch was subsequently featured in many other magazines. In 1946, Warner Brothers Studio made a movie short about the White Horse Ranch.

In 1952, Warner Brothers again visited the White Horse Ranch to make a movie short. The two movies, "Ride a White Horse" and "Ranch in White", are at the Lincoln, Nebraska State Archives and can still be viewed on video there.

Many times the horses the Thompson's used in the entertainment industry were the same horses used a few years later at the White Horse training and riding school. Some horses, such as the mare, "Snow White", served as multi-purpose horses. Snow White was a fine jumper and very fast runner. She once raced a stock car on a 1/4 mile track…and won! She knew a variety of tricks, was used as a driving horse and in a pinch could be hitched to a plow. She was extremely gentle and very popular with the training school's students. She lived into her 30s.

In 1963 Caleb Thompson died and Ruth had to close down the ranch to tourism and sell the herd. She placed breeding stock horses with friends and white horse breeders to preserve the bloodlines of Old King's descendents. But she continued the studbook and in 1970 she re-incorporated the registry as the American Albino Association, Incorporated, in Oregon. At that time the corporation began registering Creme horses in their own registry because there needed to be a better division between white horses and cream colored horses. There seemed to be confusion as to what a true white color looked like. Thus there began two distinct divisions, the American White Horse Registry and the American Creme Horse Registry, although the "breed" was still registered with the American Albino Association. Ruth was now registering horses internationally.

Carley and Dean Daugherty then became involved in the White Horse Ranch. Carley had a history with the Thompson's. As a little girl, she had ridden with the White Horse Troupe. When she was 10 years old she rode and showed the Thompson's highly schooled stallion, "Wings", an easy horse even for this young girl to handle. Wings knew an awesome number of tricks and Carley put him through his trick routine for the visitors at the ranch. She used to ride him bareback every morning to bring in the herd of show mares.

The Thompson's had been Carley's legal guardians for three years and had remained her close friends years later. In 1988 Ruth contracted with Carley and Dean Daugherty to start restoration of the White Horse Ranch which they did the following year.

In 1990 Ruth died and the White Horse Ranch was listed on the National Register of Historic Places. In 1994 the board of directors of the International American Albino Association voted to declare the American White Horse and American Creme Horse two specific breeds and to promote them as such rather than using the American Albino term. Formerly they had been promoted as two "types" within the same breed. But neither the White nor the Creme Horses were true genetic albinos, just light complected.

Today the Daugherty's continue the tradition of the Thompson's involvement with children for ranch activities, including 4-H day camps, range judging for local schools, and of coarse the White and Creme Horse herd activities. They also participate in trained horse trick routines, jumping routines and a Roman team.

Breed Basics:

The "dominant white" and the "dilute cream" are not alike in their genetic makeup; nor when reproducing is there any similarity in the effect color genes have on the foals produced. The Whites are mostly descendants of Old King, but only a few Cremes can trace their roots to him. Most Cremes in fact are from registered Quarter Horse stock.

American White Horse:

The original American White Horse had Old King as a foundation sire, although the studbook is now open to other unrelated horses of like color. The American White Horse is a color breed. Individuals within the breed will vary in conformation and type according to their background breeding. For example, one from Arab bloodlines will differ greatly from one with Quarter Horse breeding. The main requirement thus is its coloring. An American White must have a uniformly, true-white coat and pink skin. No mottling of skin such as is common to Appaloosas is acceptable. Since some horses develop a few, tiny black spots on their skin in the area of body extremities as they age, these are permitted.

Eye color will vary. White horses have been known to have the following colors of eyes: dark brown, dark blue-black, blue, very pale blue and multi-colored (sometimes called parti-eyed). They don't have hazel eyes, only the Cremes do.

White horses are born white and do not change color. The whiteness of their coats is comparable to the snow white on a domestic rabbit or the white areas on a Pinto horse. White horses should not be confused with cream colored horses. White horses come from a different, dominant white gene where a cream colored horse comes from a dilute gene.

White horses will reproduce color as follows: a white horse bred to a colored horse has 50% chance of producing a white or a colored foal. A white bred to a white horse will produce a 75% crop of white foals. The off colored foals of a white horse can be any color with chestnuts seeming to be in the higher percentage.

The American White Horse is bred not only for white coloring but for a good disposition. The breed as a whole is known to be sociable and intelligent with a willing attitude when properly treated. The American White Horse does not go blind or deaf and is not loco, as many preconceived stories have stated.

When given proper care, American Whites live long and useful lives. Many have lived into their late 20s and early 30s. Being pink-skinned does require some extra precautions, however. As with the Cremes, they can be pastured outside, but do better if they can be in the shade during midday when ultra violet rays from the sun are at their strongest, from 10:00 a.m. to 4:00 p.m. Otherwise they should be handled just like any other horse of color.

White Horses are renowned for their temperament and trainability. Carley owned and showed a gelding called "White Mystery", who was named 1963 Champion Gelding for the White Horse breed. He knew a variety of tricks, was a great jumper and also was used in Roman teams. When Carley was 9 years old, she taught him to jump with his eyes covered with a blindfold! That same summer he was introduced to a saddle for the first time and one week later she used him for saddle trick riding while on tour. She could ride him bareback with no bridle in performances where he jumped a four foot hurdle. He was so gentle that children learned to ride on him. If he was caught napping in the pasture, children would climb on his back and he would get up and take them for a quiet trip around the pasture. One winter it snowed so badly that he had to haul a makeshift travois with a 50 gallon barrel full of water up a hill to the horse pasture, never having been in harness before. He made five more trips up the hill with five more barrels without hesitation or trouble.

White Horses are beautiful and stunning to watch. In 2003, an American White, "TWHR Westwind Paris", owned by Don and Jo Ann

Anderson of Texas, was cast in two movies. He was in *Secondhand Lions* along with four other American White Horses also owned by the Andersons. Paris was featured as the main horse. He was also cast in the movie, *The Alamo*, starring Dennis Quaid as General Sam Houston and again in the movie *Rain*.

The American White Horse has proven to be very versatile and has been used for parades, shows, movies, driving, jumping, gymkhana competition, western and English pleasure, youth camp activities and handicapped riding programs. It was bred for gentleness, intelligence, beauty and color. Its disposition is excellent. Physically the American White has proven to be sound, strong and with good straight legs. It has no diseases or weaknesses particular to the breed.

American Creme Horse:

The American Creme Horse Registry was founded by Ruth Thompson in 1980 to be a registry for cream colored horses with pink skin. As such, the American Creme Horse Breed is a color breed. Horses from all major breeds which qualify by color may be registered. If the horse is the product of registered parents, its pedigree is preserved and its bloodline recognized on its registration certificate, such as Quarter Horse, Arab, etc., as long as documentation of its breeding is provided. In the event the horse has no proof of its breeding, it may be registered by color without its lineage and it would be identified as to type (stock horse, show horse, etc.).

Cremes are mostly produced by horses other than whites: usually buckskins, palominos, duns, or other cream colored horses.

Creme Horses must have pink skin. Its skin may be a darker shade, more like a tan, but must not be gray or black. It cannot have mottling on the skin such as is common with Appaloosas. It cannot have spots such as what Paints and Appaloosas have, but must have a solid cream colored coat which may vary from off white to fairly rich cream, but lighter than palomino. The mane and tail may vary from white to cream to a rich russet coloring. Cremes descending from buckskins or duns may have darker cream points on their lower legs. White markings such as blazes, stars and socks are permitted.

The dilute (recessive) gene carried in a palomino, buckskin or dun colored horse is what causes cream color. If two of these "dilute" colored horses are bred to each other, and the dilute gene from both of these parents is passed on, it produces the cream color. Cream color is ONLY produced from two dilute gene carrying parents. A White horse can ONLY produce a Creme IF it is bred to a Creme and IF the White horse carries the dilute gene. This would only happen if one of the White horse's parents or ancestors was also a dilute gene carrier and it had received that gene.

As an example, Carley had a White, blue-eyed mare. She was bred to a White, dark-eyed stud whose dam was a cremello. They produced a Creme foal. Why? Apparently the blue-eyed mare (note eye color) had an ancestor with the dilute gene. She had to donate a dilute gene along with the stallion in order for the foal to be a Creme horse.

Carley talked to Dr. D. Phillip Sponenberg, author of "Horse Color" and "Equine Color Genetics". She specifically asked him how these two White Horses produced the Creme foal. She told him the stallion's dam was a Creme, but several generations of Whites were behind the dam. He suggested the blue eyes indicated a recessive gene in the dam's background, too, thus the Creme foal.

Creme eyes can be any color, but most commonly they are pale blue or very pale amber. They can have hazel eyes, which Whites never have. Cremes never have truly pink or red eyes, but the iris may be such a pale blue that it appears nearly white and the pupil of the eye will reflect pink. However this is reflected lighting, not eye coloring.

Creme horses vary in coloring and may be born nearly white or a rich cream which looks almost gold, similar to a palomino. Cremes may lighten to almost white as they age, or in rare cases may become a darker cream color. Their skin will remain light, however. The cream color is caused by a recessive gene meaning that a Creme bred to a horse of color will dilute the color; i.e. a Creme bred to a bay will produce a buckskin and a Creme bred to a chestnut will produce a palomino. A Creme bred to another Creme will produce 100% cream colored foals.

American Cremes have proven to be of good disposition, are intelligent, and perform well according to each horse's particular type. They have been used as jumpers, trail horses, gymkhana competition horses, barrel racers, driving and ranch horses as well as pleasure and show horses.

The American Creme has a normal life expectancy if given appropriate care. However, like the Whites, their lighter skin is somewhat more sensitive to sun exposure than a dark skinned horse and it is recommended that shade be provided for them from the midday sun – from the hours of 10:00 a.m. to 4:00 p.m. This time period may need to be extended for desert or tropical regions. Sun sensitivity varies in different individuals, but only affects the areas where there is no hair covering, such as the end of muzzles and in some horses, the area around eyelids.

Recent research has proven that the light colored hooves of Cremes are just as healthy and durable as dark colored hooves.

Cremes don't go blind or deaf or become loco. Some horse people have stated differently, but it only shows lack of understanding this unique horse.

Appaloosa

Wherever there are horses, chances are there's an Appaloosa among them. That's because the word is out about this beautiful horse. Not only are Appaloosas different from any other breed, each Appaloosa is different from any other. It is the handprint of color, what Appaloosa enthusiasts call "the chrome", that is so much a part of the breed's mystique. People who own Appaloosas appreciate the difference.

But there is more to the Appaloosa than meets the eye. Breeders have worked hard to preserve the Appaloosa's special characteristics. Further,

by selecting for top performance and conformation traits, they have developed a truly extraordinary horse. The Appaloosa proves its remarkable talents repeatedly. It competes in the upper echelons of virtually every sport imaginable—from cutting to combined training, reining to racing. And while the Appaloosa more than lives up to the athletic demands of today's serious competitions, no other breed can duplicate the quality and characteristics that make the Appaloosa unique. It is the horse Nature destined to be different.

History:

The Appaloosa has a bold and colorful heritage, originating some 20,000 years ago. Its coat pattern fascinated man since the first hunters recorded its spotted image on their cave walls in what is now France. The peoples of Europe and Asia coveted spotted mounts, wars were fought over and with them, and they were often presented as gifts to the highest rulers. They were worshiped in Asia and were prized status mounts of Spanish explorers, Indians, and western settlers. Their appearance and unique qualities earned them special recognition. Legends abound about the power, tragedy, and courage of spotted horses, from Persia's Rustam and his spotted mount "Rukush", to the Blood Sweating Horses of China (spotted horses), to the story of the Ghostwind Stallions told by a Native American.

Though its ancestry can be traced back to earliest recorded time, it is in the American melting pot that the spotted horse established itself as a true breed. The Nez Perce Indians of the inland Northwest deserve much of the credit. They were renowned horsemen and the only Native

Americans known to selectively breed their horses. With their help, the true beginnings of the Appaloosa as a distinct breed began. They were documented to have had several thousand head of fleet, well formed horses, with over half estimated to be Appaloosa spotted. These they developed by their own superior breeding methods and the results were compared to the finest colonial horses in Virginia by Lewis and Clark.

Up until the association of the spotted horses with the Nez Perce, the horses with color patterns went by different names. Although the Nez Perce never called their horses "Appaloosas", the breed's name comes from "the Palouse", the region of eastern Washington and northern Idaho where the horses were known to be plentiful. White settlers first described the colorful mounts as "a Palouse horse", which was soon slurred to "A Palousey". The term "Appaloosa" is thought to have developed by this slurring.

The Nez Perce desired only the strongest, fastest and most sure-footed of mounts and used only the best animals to build their herds. Within their numbers arose a population of horses so distinctive as to inspire early American explorer Meriwether Lewis to describe them in his journal entry, dated February 15, 1806: "Their horses appear to be of an excellent race; they are lofty, elegant (sic) formed, active and durable. Some of those horses are pided (sic) with large spots of white irregularly scattered and intermixed with black, brown, bey (sic) or some other dark color…"

However, the influx of white settlers to the Northwest changed the Nez Perce's destiny and nearly destroyed the legacy of their horse breeding efforts. When they rebelled against the treaties being imposed upon them, the Nez Perce War of 1877 ensued and they were driven from their homeland by the U.S. Army. The Appaloosa helped them elude the U.S. Cavalry for several months as they traversed over 1,300 miles of rugged mountain terrain. When Chief Joseph finally surrendered in Montana, the Nez Perce were forced to relinquish their horses. The Army quickly disbanded the Indians and their fine horses were dispersed far and wide to soldiers, farmers, army Indian scouts, and even circuses. In their jealous embarrassment over being outsmarted and eluded for so long, the Army even ordered many of the Nez Perce horses destroyed.

Soon the characteristics so prized by the Indians were being lost or severely diluted due to indiscriminate breeding and the Appaloosa nearly disappeared. Some that escaped the Army or that were left behind in the Nez Perce homelands joined the herds of wild horses that roamed the plains. There were also non-treaty Nez Perce peoples from related tribes which quietly kept on breeding their treasured horses in small communities throughout the Pacific Northwest.

Horse breeders started gathering what they could of the spotted horses that were fast disappearing. They began the arduous task of preserving and re-creating the animal that had taken the Nez Perce hundreds of years to refine and that the Army had scattered to the corners of the country in just a few decades. They used the remnant Appaloosa stock that was bred to each other and it was suggested to bring in Arabian blood where necessary and possible to refine and return the Appaloosa to its former glory.

The Appaloosa Horse Club:

It was Claude Thompson, a wheat farmer from Moro, Oregon, who realized the importance of preserving the spotted horse breed. He established the Appaloosa Horse Club in 1938 to promote and restore the Appaloosa's position in the horse world. In so doing, the colorful breed began its return from the brink of obscurity. Appaloosa was the name officially adopted when the Appaloosa Horse Club (ApHC) was formed.

Those early years were a period of slow growth for the fledgling registry. The country was immersed in World War II, but after the war ended, their growth changed. In 1947, Claude Thompson appointed 23 year old George Hatley as Executive Secretary. Hatley took the shoebox containing the Appaloosa Horse Club's records to Moscow, Idaho. There were 200 registered horses and 100 Appaloosa Horse Club members.

The Appaloosa Horse Club quickly outgrew its shoebox!

Today the Appaloosa Horse Club is a major internationally recognized official breed registry. Well over a half million Appaloosas have been entered into the rolls. It sponsors annual National and World Shows and more than 100 regional clubs. With 16 percent of registrations stemming from the international market, ApHC recognizes 16 international affiliates.

THE BREED:

For most people, the Appaloosa is the easiest horse to identify due to its distinctive coat patterns, which come in an infinite variety—from no spots at all, to more spots than can be counted. Some Appaloosas look like a covering of fresh-fallen snow over their backs, loins and hips. Some Appaloosas are described as "leopards," with Dalmatian-like spots. Some are Appaloosa roans. This is one of the exciting aspects of the Appaloosa – its enormous range of coat colors and patterns. Any number of spotting combinations is possible. The variety and unpredictability is especially intriguing to breeders.

However, Appaloosa patterns should not be confused with overo and tobiano markings of Pintos or Paints. Pinto markings tend to be larger and a different shape and placement than those of the Appaloosa. Appaloosa spots and spotting patterns can change over the course of their lifetime, but Paint/Pinto spots remain the same.

There are three other visible traits beside coat patterns that contribute to the Appaloosa's unique appearance. One is the irregularly pigmented or "mottled" skin which is most apparent around the horse's muzzle, eyes and genitals. Also, many Appaloosas have a distinctly human-looking eye due to the white sclera surrounding the iris, the dark-colored center portion of the eye. Many Appaloosas have vertical stripes on their hooves in the absence of white leg markings. An Appaloosa need only have mottled skin and one other visible characteristic for regular registration status.

However, not every Appaloosa is blessed with these easily identifiable traits. Recognizing this quirk of nature, the Appaloosa Horse Club also accepts non-characteristic horses for registration. Non-characteristic Appaloosas can be used for breeding purposes: however, in order to be shown at Appaloosa Horse Club approved events, they must undergo parentage verification and inspection. Breeders may mate Appaloosa to Appaloosa, or crossbreed to registered American Quarter Horses, Thoroughbreds or Arabians.

Registration:

The four identifiable characteristics are: coat pattern, mottled skin, white sclera, and striped hooves. In order to receive regular registration, a horse must have a recognizable coat pattern OR mottled skin and one other characteristic. Horses which receive regular registration are issued numbers (no letters precede the number, but the # sign will). Those not displaying a coat pattern or mottled skin and one other characteristic will be classified as non-characteristic and their registration numbers will be preceded by the letter "N". They can't compete in ApHC approved events.

Stallions are registered either with the ApHC or with an approved breed registry which participates in the ApHC breeding program. They must also be DNA typed.

Mottled or Parti-colored skin:

This characteristic is unique to the Appaloosa horse. Therefore mottled skin is a basic and decisive indicator of an Appaloosa. Mottled skin is different from commonly found pink (flesh colored or non-pigmented) skin in that it normally contains dark areas of pigmented skin within its area. The result is a speckled or blotchy pattern of pigmented and non-pigmented skin.

If a horse has mottled skin, it may be found in several places in addition to the muzzle and eye areas. Many breeds have specks of non-pigmented skin which should not be confused with the Appaloosa's mottled skin.

White Sclera:

The sclera is the area of the eye which encircles the cornea – the colored or pigmented portion. The white of the human eye is an example. All horses have sclera but the Appaloosa's is white and usually more readily visible than other breeds. All horses can show white around the eye if it is rolled back, up or down, or if the eyelid is lifted. Readily visible white sclera is a distinctive Appaloosa characteristic provided it is not in combination with a large white face marking, such as a bald face.

Striped Hooves:

Many Appaloosas will have bold and clearly defined vertically light or dark striped hooves. But all striped hooves do not necessarily distinguish Appaloosas from non-Appaloosas. Further Appaloosa characteristics in these situations apply.

Base Coat Colors:

The Appaloosa Horse Club recognizes the following 13 base colors: bay, dark bay or brown, black, white (snow white with pink or light-colored hide. Some horses have a white body with dark spots over part or all of their bodies, sometimes referred to as "leopard"), buckskin, chestnut, dun, gray, grulla, palomino, red roan, bay roan and blue roan.

"Varnish marks" are roan markings where the darker color appears in other areas on the body such as behind the elbows, across the flanks, etc.

It's not always easy to predict the color a grown horse will be from the shade it appears to be as a foal. Most foals are born with lighter-colored coats than they will have when they shed later, with the exception of gray horses.

Coat Patterns:

A remarkable aspect of the Appaloosa is the myriad of color and pattern combinations they can exhibit. Appaloosa patterns are highly variable and there are many which may not fit into specific categories easily. Seven common terms used to describe coat patterns are:

Blanket –solid white commonly over, but not limited to, the hip with a contrasting base color.

Spots – white or dark spots over part or all of the body.

Blanket with spots – white blanket which has dark spots within the white.

Roan – lighter color on parts of the head, over the back, loin and hips. Darker areas may appear along the face frontal bones, above the eye, and legs, stifle, point of the hip and behind the elbow.

Roan Blanket – roan colored body with a blanket over, but not limited to, the hip area.

Roan Blanket With Spots – roan blanket which has white and/or dark spots within the roan area.

Solid – base colored with no contrasting color in the form of an Appaloosa coat pattern.

Disqualifications: No horse shall be registered that has:
- artificial characteristics or coat patterns
- draft, pony, albino, Pinto, or Paint breeding
- continuous leg marking(s) which exceed a line around the throat latch and behind the ears and/or…
- white marking(s) on the body which are continuous, uninterrupted, longer than six inches and separate from an Appaloosa coat pattern, if a pattern is present, and which marking(s) do not blend into the base color.
- one parent that is registered with non-breeding stock papers with an approved breed association
- is less than 14 hands after they are 5 years old or older

The Active Appaloosa:

The Appaloosa's color, versatility, willing temperament and athletic ability make it a popular choice for a number of activities. ApHC classes include roping, jumping, gymkhana events, halter, saddle seat pleasure, and heritage classes. They also excel at endurance riding. Many competitive endurance riders have found that the Appaloosa has the toughness, resilience, heart and stamina to travel 25, 50, even 100 miles in a single day.

Appaloosas can also race. Fierce competition within the Appaloosa racing industry produces some of the fastest horses in the world. The breed is recognized as the "middle-distance runner." They compete at distances from 220 yards to eight furlongs. Appaloosa athletes continue to set and break all-breed speed records. Annually, more than 450 Appaloosa races pay in excess of $2,000,000.

Owners have the luxury of breeding for specific conformation and performance traits—whether selecting an agile western stock horse or a grand dressage horse. The Appaloosa's trustworthy disposition and willingness to please makes it ideal for all ages.

The International Colored Appaloosa Association, Incorporated:

Breeders in this association are normally referred to as "foundation breeders" and the horses referred to as "foundation type" or "blood-breed Appaloosas".

Appaloosa research released in 1994 estimated there were less than 3,000 living registered Appaloosas whose parents, grandparents, and great-grandparents were all registered as Appaloosa rather than another breed registry. Thus the International Colored Appaloosa Association, Incorporated (ICAA), was conceived in 1991 by several concerned, long-time Appaloosa breeders and owners for the purpose of saving and restoring the Appaloosa as a breed. The ICAA's intent was to preserve the blood-breed ("pure") Appaloosa and its heritage as well as to promote this versatile and athletic breed throughout the world. Their founding Board of Trustees states: "Today the Appaloosa breed… teeters on the brink of extinction." Therefore, they wanted to "return the Appaloosa breed registry"…to the "purity of the breed".

The ICAA is the first Appaloosa blood-breed registry with its books closed to other breeds and its registry with classifications based only on pedigrees. The purpose is to preserve the pure equine breed of Appaloosa from undesirable introductions or influences of other equine breeds. It doesn't accept cross-bred horses for registration except geldings showing the Appaloosa coat pattern. The goal is to save the original Appaloosa with all of its complete breed traits and to restore it to the admiration and respect it deserves. The ICAA promotes the return of the breed to its original conformation, which would otherwise experience certain extinction. The concept is that there is no such thing as "too much Appaloosa" in a pedigree.

Proper breeding practices show that the concept of the big-footed, draft-type "pure Appaloosa" is inaccurate and not what the original Appaloosa appeared to be. Some traits to look for in an ICAA registered

International Colored Appaloosa

Appaloosa are a neck set a bit higher than others due to their use in chasing buffalo, which required seeing higher and moving faster. (A horse sees farther with the head held high rather than low, which is its area of close vision.) Also, a strong back and deep heart girth, plus good bone on the legs are desirable.

Appaloosas have a variety of head types. Decent-sized ears (for better hearing) and a throatlatch that allows easier breathing are expected. Small feet, small ears, and thin necks are not ideal characteristics for this versatile using horse. Appaloosas are also renowned for their intelligence.

Currently it is impossible to be more definitive as to what the conformation of a blood-breed Appaloosa should look like. There are no Appaloosas with eight generations of purebred foundation stock to use as a guide.

Soundness and the physical/mental abilities to perform the skills asked of it are at the top of the list in importance. This is achieved by concentrating the Appaloosa gene pool toward a purebred standard. With each additional generation, the ICAA Appaloosa has a higher concentration of Appaloosa genetics behind it.

The true Appaloosa is athletic, versatile and has a family-orientated disposition. It is a distinct breed of horse, reminiscent in construction, attitude and ability of the highly-regarded Appaloosa war and buffalo horses of the Northwest... the world's best rough country stock horse.

AraAppaloosa

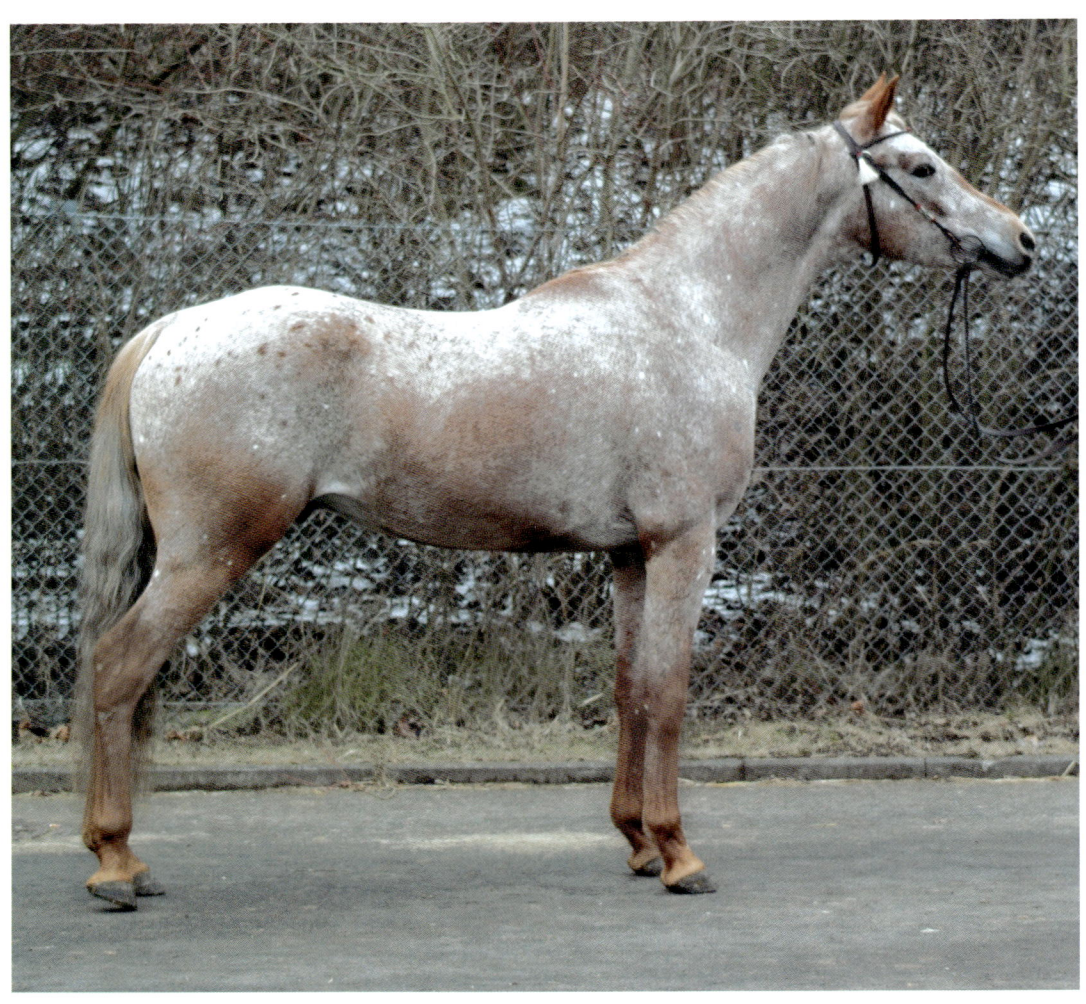

The AraAppaloosa is a more refined Appaloosa of extraordinary quality; one with color, elegance, performance ability, soundness, stamina, and endurance. It is durable, intelligent and has a great disposition, yet is capable of great spirit.

It combines the color, personality and good temperament of the Foundation bred Appaloosa with the bloodlines and spotlines of the Arabian, which traces back in an unbroken line to the spotted horse of the ages.

FOUNDATION APPALOOSA

At its inception many years ago, the ApHC (Appaloosa Horse Club) certified the first 4,932 horses to attain permanent status through inspections as Foundation Appaloosas, and gave these horses the "F" prefix before their number. These first horses were considered to be the correct type by the inspectors, who also observed to see if they produced foals of color. Many of these Foundation horses were refined and excellent examples of well-planned breeding programs. They also carried a high percentage of oriental blood along with the Appaloosa blood tracing to the horses of the Nez Perce, Palouse, and other Indian tribes.

Many of the Foundation horses had distinct characteristics and qualities worth passing on. Also, many were genotypes capable of reproducing their color and other attributes several generations down the line.

Some of the Foundation sire lines known for producing high performance offspring with refinement are: Apache, Arab Towsirah Alkhar, Bear Paw, Mansfield's Comanche, Freels Chico, Patches, Patchy, Peter K, Red Eagle, Sundance and Toby.

The AraAppaloosa Foundation Breeders' International (AAFBI) strives to preserve these Foundation lines and welcomes Foundation bred Appaloosas carrying three or more generations of Appaloosa bloodlines and/or oriental blood.

SPOTLINE ARABIAN HORSE

Spotline purebred Arabian horses trace to parti-colored purebreds of the Arabian Desert which had white spots or spotting along with white sclera, some mottled skin and/or striped hooves — characteristics that make the Appaloosa breed unique. Although some experts do not concede that these characteristics on an Arabian horse are a result of Appaloosa genes, it is generally agreed that genes producing such color features would certainly compliment the Appaloosa.

Many examples of spotted Arabians appear in ancient paintings. The earliest records of Arabian horses are in Egyptian and Middle East artwork which frequently depicted them with spotted-like sections on their coats. Because the Arabian has been a distinct breed type for thousands of years, this is an important statement about what the foundation of the breed looked like. There is no denying that the original Arabian had spotting. From this it can be ascertained that spotted Arabians today denote the oldest and purest lines.

There are myriad examples of these markings that appear in breeding logs and early photos as well as on many modern Arabian horses. Most commonly they are called the "bloody shoulder" and "kellogg spotted". "Bloody shoulder" refers to red hairs in splotchy patches on one or both shoulders that are seen mostly in gray Arabs. Sometimes it's not visible unless the horse is sweating or being bathed.

"Kellogg spotted" comes from the line of Arabians that were imported by Mr. W.K. Kellogg (famous for Kellogg cereals) to America earlier in the 20th century. These were superb horses from the famous Crabbet Stud in England. A couple of his horses had spots and passed them on to their offspring. The same can be said of the other Crabbet and Babson lines.

Also, Andalusians, Lipizzaners and other Arabians commonly have mottled skin and striped hooves. Thoroughbreds (which are descendants of oriental horses such as the Byerly Turk, the Darley Arabian and Godolphin Barb), often display spots on their hindquarters. The most notable spotted Thoroughbreds are "Bend Or", "Man O'War", and "Candy Spots"

It has been proven that the Arabian spotlines complement and strengthen the Appaloosa spotting gene. The roaning gene (not to be confused with the undesirable graying gene) is also useful in attaining the optimum color in the AraAppaloosa horse.

HISTORY

In 1877, when the Nez Perce failed to escape the U.S. Army, the Calvary confiscated and destroyed many of the Nez Perce Appaloosas. What horses survived the Army were crossed with draft animals in an attempt to destroy two centuries of Nez Perce selective breeding.

Claude J. Thompson of Moro, Oregon, founder of the Appaloosa Horse Club, infused the blood of European-bred horses into his native Appaloosas as a way to improve their structure, which the army's draft crosses had altered. He made great strides in restoring the Nez Perce Indian type of Appaloosa by redeveloping its light body with Arabian breeding. He commented, "Having some knowledge of the Arabian horse, and knowing that most light breeds were established on Arabian blood, I made a trip to California where I selected and brought home a pure Arabian, "#922 Ferras", to refine and improve the conformation of my Appaloosa horses."

Ferras's sire and dam were two registered Arabians, "*Ferdin" and "*Rasima", who were imported from the famous Crabbet Stud of England. Claude most often crossed Ferras to mares of the Old Painter line as a way to refine their heavier frames. Seventeen of the foundation Appaloosas were sired by Ferras, whose most famous offspring was "Red Eagle" ApHC #F-209. Claude crossed Ferras with "Painter's Marvel" ApHC F-47 to produce Red Eagle. Painter's Marvel is a granddaughter of Ferras. Her sire, "Painter III" is number F-8, while her dam, "Snowflake" is F-2.

Red Eagle represents a turning point for the Appaloosa breed. He was a bay and white with bay spots and foaled in 1946. In 1951 Red Eagle won the National Champion stallion title. Later, actor John Derek bought Red Eagle for use in a western film. The movie never materialized and Derek sold Red Eagle to Thomas Clay of Caliente, Nevada. While in Nevada, Red Eagle made a name for himself by siring many great horses. Red Eagle's most prominent offspring included "American Eagle, F-1472", "Simcoes Frosty Eagle" and Hall of Fame stallion "Red Eagle's Peacock F-1476". Red Eagle sired 81 registered foals who earned several National Championships and record wins of two Bronze Medallions.

During the 1930s and 1940s, breeders of the foundation Appaloosas primarily used Arabians and Arabian crosses to refine their spotted horses. More than 30 Arabian stallions are listed as the sire of many foundation Appaloosas.

Today there is a surplus of the heavy, Quarter Horse type Appaloosa. However, the lighter types infused with Arabian or Thoroughbred blood can still be found. A small percentage of Appaloosa breeders in the United States are determined to keep Claude Thompson's dream alive.

The type of horses needed to reproduce and continue to breed in the future is the spotted horse of the past, the one the Nez Perce Indians bred for and the one that is pictured time and again in paintings and descriptions throughout history. The AraAppaloosa Foundation Breeders' International (AAFBI) is a registry that focuses on breeding horses within this heritage of the spotted horse in which the Arabian breed had a vital role. Both plans – breeding Appaloosa to Appaloosa and delving back in the lineage to incorporate the oriental/Arabian blood into a breeding program – achieves the same goal: to produce the authentic Appaloosa of the past – the AraAppaloosa.

Owners of valued Appaloosas with Arabian traits are attracted to the breed. In 1997, when Julie and Randy Berghammer found it difficult to get registration papers on their Appaloosa mare, they got involved with AAFBI, which had a hardship clause available for those horses that met the criteria of color and conformation. Their mare was a varnish roan with dark legs and a spot on each hip (commonly referred to as a varnish mark). She also had good conformation and a dished face. She produced a varnish colored daughter just like herself and a son who also displayed lots of color and a dished face. He was a dark chestnut with a hip blanket, spots, marbling on his barrel, and lightening marks on his legs – common Appaloosa traits.

With their beautiful Appaloosa color and obvious Arabian traits, these horses were the type that epitomized the AraAppaloosa. The Berghammers found that there were other horses with similar Appaloosa and Arabian combined traits that fit the AraAppaloosa description.

REGISTRATION

AraAppaloosa breeders work to produce horses that have excellent dispositions, superior athletic conformation, beauty and Appaloosa color. With the use of pedigrees, it can be understood what bloodlines help achieve the lofty goals of producing the perfect AraAppaloosa.

When faced with setting up a system to rate horses, the AAFBI decided to emphasize known pedigrees and mathematically determine the percentage of blood from these known sources. They also emphasize

Appaloosa color and characteristics. A breeding stock certificate is available for the solid colored horses that are produced from qualified parents.

A hardship registration is also available for those horses that have no known background. These horses must display obvious Appaloosa color and characteristics and conformation must be of the light horse type.

Registration certificates contain a prefix rating, a number and a suffix rating, such as "A 44 ap". The "A" rating means the horse has at least 1/2 Arabian blood, the 44 is the number in the registry and the "ap" means that the horse has less than 1/2 Appaloosa blood. The entire rating list is as follows:

"a" prefix – at least 1/4 Arabian blood	"ap" suffix – less than 1/2 Appaloosa blood
"A" prefix – at least 1/2 Arabian blood	"Ap" suffix – at least 1/2 Appaloosa blood
"2A" prefix – at least 3/4 Arabian blood	"2Ap" suffix – at least 3/4 Appaloosa blood

The AraAppaloosa and Foundation Breeders' International bloodline rating was established for the express purpose of helping the breeder and owner of an AraAppaloosa understand what is behind their horse's pedigree. The rating helps determine exactly how much actual Appaloosa blood and Arabian blood (which includes Thoroughbred) is in the horse being registered. The AAFBI rating is unique because it is based on the most recent 4 generations. Therefore, as a breeding program continues to focus on restoring the Nez Perce type Appaloosa, the future foals' rating can improve. (*AraAppaloosa Fact Sheet, The AAFBI Registry and Focused Restoration, by John L. Baker.*)

CHARACTERISTICS

With a height varying from 14 hands to 16 hands, the general appearance of the AraAppaloosa should fit the Registry's slogan "tough but elegant". The head must be small and refined. A dished face similar to the Arabian's is desirable but not essential. Eyes should be large and encircled with white sclera; ears should be small and curved inward like crescents.

The neck of the AraAppaloosa should have a natural arch but shouldn't be overly long. Withers are not prominent and the back is short. Although the ideal topline for an Arabian is level, the AraAppaloosa should have a slightly sloping croup and a long, sloping hip, a deep heart girth and a long underline. The overall appearance of the body should be symmetrical and athletic.

An AraAppaloosa in action is almost unsurpassable in beauty. It has the springy prance, flared nostrils, arched neck and tail carriage of the Arabian, coupled with the striking contrast of a white blanket and/or spots, striped hooves and sclera-encircled eyes.

Because of its oriental heritage, an AraAppaloosa foal is likely to inherit loud coloring. Many AraAppaloosa foals are born with little or no color, then color "out" as they mature. However, characteristics such as evident sclera, mottled skin and striped hooves should be present from birth, even if color is not. A variety of patterns exist, but the following are most common:

Spotted Blanket. Most breeders prefer this classic pattern. The foreparts of the horse are solid colored while its rump has a white "blanket" with dark spots. Blankets may vary in size and spots vary in shape from large egg-shaped spots to halo (roan edged) or teardrops.

White Blanket. The foreparts are dark and a solid white blanket covers the hindquarters. The blanket may extend to the shoulders.

Marble. The marble pattern comes with maturity. Foals that have this pattern are usually born with a solid color, then roan as they age. Marble AraAppaloosas are often white or roan with speckled spots over their hindquarters. Faces, legs and necks, however, are often a much darker color. Variations in this pattern are red, strawberry and blue roans.

Leopard. A marble may seem to be a leopard, but a true leopard AraAppaloosa is foaled snow-white with dark spots over the entire body. A "near leopard" is similar to the true leopard, except that the face and upper neck of the horse are a solid color. The legs may be dark with white lightning marks on the cannons. A "fading leopard" is born pure white with loud spots over the coat. If the sire or dam is a gray, however, the unfortunate owner is sure to see the horse's spots fade with age, resulting in a pure white horse.

Snowflake. This is a dark base color with white spots or flecks sprinkled over the entire body and neck. The snowflake AraAppaloosa is often born solid, then colors with age. There is also a pattern called a snowflake blanket in which the white spots are concentrated on the rear quarters.

Lace or Frosted Blanket. The base color of the horse is dark, with white "lace" on its rump. Foals that will later develop a spotted or white blanket are often born with a muted lace blanket. The blanket may have spots of any size, but the spots are often not as noticeable because there is less contrast.

The future of the AraAppaloosa is toward sporthorse competition. Many are in endurance racing and several have competed in 3-day eventing and dressage.

In Europe, the expanding popularity of the breed is seen in the famous stallion, "Congal's Cosmic Sky". Sky is an AraAppaloosa in Germany that passed the Körung (an advanced breeding suitability test) and is listed with the First Stud Book, which only accepts stallions of the highest perfection with unusually superb conformation and movement. This is a rare and prestigeous honor only bestowed on the very best quality stallions. Sky is the only AraAppalosa in Germany to have achieved this so far, but he initiates the excellence of the AraAppaloosa breed to Europe.

AraAppaloosas are great family horses and are quite often used in therapeutic riding programs because of their gentle nature. They make excellent all around horses, excelling at just about anything they are trained for. They are intelligent and pick up training quickly. In the show ring is where they really shine; whether in dressage, hunter/jumper, western pleasure, or saddle seat, they have the show ring presence that is breathtaking.

The AraAppaloosa projects its fine tradition of elegance. It is a rare and beautiful breed.

Arabian

Romance, antiquity and tradition are integral parts of the Arabian breed. Even more, Arabians possess unique qualities that distinguish them from other breeds of horses – their incredible beauty, endurance and intense affection and loyalty to their owners. Historical figures such as Napoleon and America's first president, George Washington, rode Arabians.

The breed is distinctive with their dished head profile, large, lustrous, wide-set eyes on a broad forehead, small, curved ears and large, efficient nostrils. Known for their deep chest, strong joints and powerful lungs, Arabians are all-around athletes who can do many things well and LOVE challenges.

Authorities are at odds about where the Arabian horse originated. There are certain arguments for the ancestral Arabian having been a wild horse in northern Syria or southern Turkey.

But about 3500 years ago in an area including the valley of the Nile and beyond, the forerunner of the breed that eventually became known as "Arabian", was first noticed when it attained a role alongside kings. Its image appeared on seal rings, stone pillars and various monuments with regularity after the 16th century B.C. This ancient horse was observed for many centuries before the word "Arab" was ever used or implied as a race of people or species of horse.

The origin of the word "Arab" is still obscure. "Arab" is a Semitic word meaning "desert" or the inhabitant thereof, with no reference to nationality. In the Koran, "a'rab" is used for Bedouins (nomadic desert dwellers).

Through the centuries the Bedouin tribes who roamed the northern desert in what is now Syria became the most esteemed horse breeders. Their horses reached a zenith of fame as the horse of the "Arabas". The harsh desert environment ensured that only the strongest and keenest horses survived, and they passed down many of their resilient physical characteristics that distinguish Arabians to this day.

The Bedouin people bred their horses as war mounts capable of quick forays into enemy camps. The severe desert climate required these nomads to share their food, water and sometimes even their tents with

the horses. Over the millennia, Arabians developed a close affinity with man as well as a high level of intelligence.

Bedouins zealously maintained the purity of the breed. Even today, the purebred Arabian is virtually the same as that ridden in ancient Arabia. Due in part to the religious significance attached to their desert horse as well as the contribution it made to the wealth and security of the tribe, the breed flourished in near isolation. Traditions of breeding for purity were established and any mixture of foreign blood from the mountains or the cities surrounding the desert was strictly forbidden. While other desert type breeds developed in North Africa and the periphery of the Great Desert, they were definitely not of the same blood as Arabians and were disdained by the proud Bedouins. Arabian mares evolved as their most treasured possession. To the Bedouin, no greater gift could be given than an Arabian mare.

The value placed upon the mare led inevitably to the tracing of any family of the Arabian horse through the dam. The only requirement of the sire was that he be pure. If his dam was a "celebrated" mare of a great mare family, so much the better. Mare families, or strains, were named often according to the tribe or sheik who bred them.

The Bedouin valued purity in a strain of horses above all else, and many tribes owned only one main strain of horse. The five basic families of the breed include Kehilan, Seglawi, Abeyan, Hamdani and Hadban. Other less "choice" strains include Managhi, Jilfan, Shuwayman, and Dahman. Sub-strains developed in each main strain and were named after a celebrated mare or Sheik that formed a substantial branch within the main strain.

A great story of courage, endurance, or speed always accompanied the recitation of the genealogy of the sub-strain, such as the great Kehilet al Krush, the Kehilet Jellabiyat and the Seglawi of Ibn Jedran. Each of these mares carried with them stories of great battles and intrigue. Their daughters were sought after by the most powerful kings, but often remained unattainable. Daughters and granddaughters of these fabled mares changed hands through theft, bribery and deceit. If any of their descendants were sold, the prices were legendary.

Each strain, when bred pure, developed characteristics that could be recognized and identified. The **Kehilan** strain was noted for depth of chest and masculine power and size. The average pure Kehilan stood up to 15 hands. Their heads were short with broad foreheads and great width in the jowls. Most common colors were gray and chestnut.

The **Seglawi** was known for refinement and almost feminine elegance. This strain was more likely to be fast rather than having great endurance. Seglawi horses displayed fine bone and longer faces and necks than the Kehilan. The average height for a Seglawi would be 14.2 hands and the most common color was bay.

The **Abeyan** strain was very similar to the Seglawi. They tended to be refined. The pure Abeyan would often have a longer back than a typical Arabian. They were small horses, seldom above 14.2 hands, commonly gray and carried more white markings than other strains.

Hamdani horses were often considered plain with an athletic, if somewhat masculine, large boned build. Their heads were more often straight in profile, lacking an extreme "jibbah" (bulging forehead). The Hamdani was one of the largest, standing as much as 15.2 hands. Common colors were gray and bay.

The **Hadban** strain was a smaller version of the Hamdani, sharing several traits, including big bone and muscular build. They were also known for possessing an extremely gentle nature. The average height of a Hadban was 14.3 hands and the primary color was brown or bay with few, if any, white markings.

Bred in the desert, the Bedouins' remarkable horses had evolved like finely tempered steel into the swift, elegant, graceful and magnificent war horses by whose means the Arabs shook the civilized world. Over the centuries, they zealously maintained the purity of the breed. Because of limited resources, their breeding practices were extremely selective, which eventually established the Arabian as a prized possession throughout the world.

EARLY BREED EXPANSION:

Europe had developed bigger horses through the Dark Ages to carry a knight in armor and their lighter horses were from the pony breeds. They had nothing to compare with the smaller, faster horses of the invading Turks. Though very few Arabians accompanied these invaders, an interest in "Eastern" horses grew. To own such a horse would not only allow for the improvement of local stock, but would endow the fortunate man with incredible prestige. Such a horse in the stable would compare in value to owning a great artistic painting. Europeans of means, primarily royalty, went to great lengths to acquire fabled Eastern horses.

So European horses soon had an infusion of Arabian blood, especially as a result of the Christian Crusaders returning from the East between 1099 A.D. and 1249 A.D. With the invention of firearms, the heavily armored knight lost his importance and during the 16th century the handy, light and speedy horses became more in demand. Eventually, people of the western world began looking to the people of the East for Arabian stock.

As the world slowly shrank due to increasing travel abroad, the Turkish rulers of the Ottoman Empire began to send gifts of Arabian horses to European heads of state. Thus a revolution in horse breeding occurred when three Arabian stallions were imported to England: "The Godolphin Arabian" (sometimes called "Barb") imported in 1730, "The Byerley Turk" (1683) and the "Darley Arabian" (1703). These three "Eastern" stallions formed the foundation upon which a new breed, the Thoroughbred, was to be built. Today 93% of all modern Thoroughbreds can be traced to one or more of the three sires. By direct infusion and through the blood of the Thoroughbred, the Arabian has contributed in some degree to all our light breeds of horses.

The Arabian horse also made inroads into other parts of Europe and even farther east. In France, the Arabian helped to make the famous Percheron and the Limousin (to this day bred as the French Anglo-Arab). Two Arabians were used as foundation sires of the Lipizzaner. In East Prussia the Arabian was used to create the Trakehner and in Russia it contributed extensively to the development of many breeds, including the Orloff Trotter.

In the 1800s, travelers of the Victorian era became enamored with the horse of the desert and significant Arabian stud farms were founded throughout Europe. The royal families of Poland established notable Arabian studs, as did the kings of Germany and other European nations.

As a result of Lady Anne Blunt and Wilfred Blunt's historical sojourns into the desert to obtain Egyptian and desert stock, the world-famous Crabbet Arabian Stud was founded in England. This stud farm eventually provided foundation horses for many countries, including Russia, Poland, Australia, North and South America and Egypt.

IN THE U.S.A.

America was built by utilizing horsepower and colonists there also were quick to realize the value of Arabian stock. Nathan Harrison of Virginia imported the first Arabian stallion in 1725. This horse reportedly sired 300 foals from grade mares.

In 1877, General Ulysses S. Grant visited Abdul Hamid II, His Imperial Majesty the Sultan of Turkey. There he was presented with two stallions from the Sultan's stable, "Leopard" and "Lindentree". Leopard was later given to Randolph Huntington who subsequently imported two mares and two stallions in 1888 from England. This program, limited as it was, must be considered the first purebred Arabian breeding program in the United States.

The Chicago World's Fair held in 1893 drew widespread public atten-

tion and had an important influence upon the Arabian horse in America. While every country in the world was invited to participate, Turkey chose to exhibit 45 Arabian horses in a "wild Eastern" exhibition. Among the imported Arabians shown were the mare "Nejdme" and the stallion, "Obeyran", who became foundation animals No.1 and No.2 in the Arabian Stud Book of America. Several years later, two other mares and one stallion were also registered. Many breeding farms today have horses whose pedigrees trace to these 19th century Arabians.

One of the most significant imports occurred in 1906 when Homer Davenport received permission from the Sultan of Turkey to export Arabian horses. Davenport, with the backing of then President Theodore Roosevelt, imported 27 horses that became the foundation of Davenport Arabians.

In 1908, the Arabian Horse Club of America was formed (now known as the Arabian Horse Registry of America) and the first stud book was published. Recognition of the Arabian Stud Book by the U.S. Department of Agriculture established the Registry as a national registry and the only one for the purebred Arabian breed. Seventy-one purebred Arabians were registered at that point.

Today the Arabian horse exists in far greater numbers outside of its land of origin than it ever did in the Great Desert. There can be no dispute, however, that the Arabian horse has proved to be, throughout recorded history, an original breed – which remains to this very day.

ARABIAN EXCELLENCE

Given that the Arabian was the original source of quality and speed and remains foremost in the field of endurance and soundness, it either directly or indirectly contributed to the formation of virtually all the modern breeds of horses. As the oldest of all the light breeds and foundation stock of most, it is unique. The Arabian is different in that it does not exist as a result of selective breeding like other modern light breeds who established a registry prior to their development. But it was a breed that had been recognized for thousands of years and had been maintained and cherished in its purity over those years as much as is humanly possible. Its influence is commanding to a remarkable degree and invariably dominates all the breeds to which it is introduced as it contributes superior qualities.

Because the Arabian was bred and reared in close contact with man from the earliest records and existed in mutual inter-dependence, it developed an unequaled ability to bond with humans. It is gentle, affectionate, and familiar, almost to the point of being troublesome. Foals, for example, have no fear of man, and are usually indifferent to sudden noises. The Arabian gentleness and tractability, while originally the effect of education, is now inherited, as foals bred in a foreign environment are demonstrating.

Even today the purebred Arabian is virtually the same as that ridden in ancient Arabia. Its ancient traits enable him to excel at many versatile activities, displaying athletic talents in disciplines from English to western riding.

Its stamina is legendary as an endurance horse. It is considered the best breed for covering long distances that would exhaust other breeds – while still having energy for more. It is superior at consistently winning competitive trail and endurance rides: the top prizes at endurance events almost always go to riders of Arabians. As an endurance horse, the Arabian has no equal and is the undisputed champion.

The loyal, willing nature of the Arabian suits itself as the perfect family horse for recreational riding. His easy going temperament and smooth gaits are terrific for beginning riders. His affectionate personality makes him a great horse for children. Growing up around Arabian horses is an experience that lasts a lifetime.

In the show ring the Arabian is exceptional, well known for his balance and agility. Combined with his high intelligence and skillful footwork, he is more than capable in driving and reining events. The Arabian competes in more than 400 all Arabian shows as well as in numerous open shows around the U.S. and Canada. For speed, agility and gracefulness, there is nothing like an Arabian.

As the original racehorse, he is becoming increasingly more popular competing at racetracks throughout the country. Arabians race distances similar to Thoroughbreds and more than 700 all Arabian races are held throughout the U.S. annually.

Today's Arabian prices are comparable with other popular breeds and now excellent Arabian horses are accessible to a broad base of horse enthusiasts. With more living Arabian horses in the United States than in all the other countries in the world combined, America has some of the best horses and breeding farms from which to choose.

BREED CHARACTERISTICS:

The Arabian is a blood breed based on pedigree. Horses that look like Arabians cannot be introduced into the breed. All registered Arabians are validated via DNA to both parents who must be registered purebred Arabians. A horse conceived in the U.S. or Mexico and sired by an eligible stallion, and born in the U.S. or Mexico, can be registered with the Arabian Horse Association.

An Arabian can most readily be identified by its finely chiseled head with a dished face, long arching neck, and high tail carriage. When looking at an Arabian, it is immediately noticed how its entire appearance exudes energy, intelligence, courage and nobility.

The Arabian is known for a well coordinated, free, easy stride with stylish, natural, balanced action. Every time an Arabian moves in its famous floating trot, he announces to the world his proud, graceful nature.

Many of the Arabian's distinct characteristics proclaim its desert heritage. Long eyelashes are designed to protect the eyes from sand while large nostrils assure easy breathing in a hot, dry climate. Likewise, the Arabian's deep chest, strong joints and good lungs guarantee its ability to carry riders across large stretches of desert.

The breed is noted for its delicate head, featuring small, tipped-in ears, a tiny "teacup" muzzle and large, expressive, dark eyes. The head has a triangular shape that diminishes rapidly to the small, fine muzzle. The lips are fine and thin. The nostrils are long, thin and delicately curled, running upward and outward. The eyes are set far apart, nearly in the middle of the head, and are large and lustrous. When the horse is aroused, they are extremely attentive.

Characterized by its fine, chiseled appearance, the Arabian head is frequently enhanced by a slight protrusion over the forehead that extends to just below the eyes. This protrusion, called the "jibbah", contributes to the Arabian's distinct appearance. The cheek bones are spread wide apart at the throat, enabling the muzzle to be drawn in without compressing the windpipe, and permitting the animal to breathe easily when running.

The eloquent Arabian head has been represented artistically for literally thousands of years. Defined, described, and judged for centuries, the shape and beauty of the Arabian head remains its most distinctive and sought after quality.

The neck is long and arched, set on high and run well back into the withers. The withers should be prominent and set well back over a long, sloping shoulder and equal in height with the croup and with a high tail set. One of the most distinguishing characteristics is its naturally high tail carriage.

In general, Arabians have a short, straight back, a deep chest, well-sprung ribs, deep girth and strong legs of thick density. They have perfect balance and symmetry.

The Arabian's skeleton is characterized by a relative shortness of skull, a slenderness of the lower jaw and a larger size of the brain case. It has fewer vertebrae in the back (usually 23 vertebras as compared to 24 in most other equine breeds) and also in the tail. The pelvic bone has a more horizontal position.

In height, the Arabian horse generally measures 14.1 to 15.1 hands at the withers, although there are horses above or below this height.

In color, Arabians are bay, gray, chestnut and black, with an occasional roan. Common markings are stars, strips, snips or blaze faces, and white feet or white stockings. There are no restrictions about the amount of white on a purebred Arabian.

Today's Arabians are also recognized for their extraordinary intelligence, stamina, and trainability. Although the most beautiful of all riding breeds, the Arabian is not just a pretty horse. He is an all-around family horse, competitive sport horse, and work horse. Trail riding, endurance riding, showing, cutting, reining, dressage, jumping, racing and even ranch work – If a rider is up to it, so is his Arabian!

Brindle

The brindle pattern consists of a watery or drippy looking striping – or sometimes just partial striping – over the body of an animal. It is more commonly seen in dogs or cattle. In horses, the pattern is extremely rare.

Many consider brindle colored horses a curiosity, which is better than the discrimination unusual colored horses received in the past. "I bought my Brindle horse just because the coat pattern was so wild and eye-catching," says Denise Charpilloz of Washington. "I have always been attracted to colorful horses. The brindle pattern is so rare people would approach me to ask, 'What IS that?' Also, people would point out my Brindle horse to each other, even if they didn't always come up and speak to me directly. More often than not, old-timers will tell me they have been breeding horses for more than 30 years and had NEVER seen anything like THAT!"

Most people say they never knew anything like it existed in horses. When Brindle horse owner J. Sharon Batteate picked up one of her Brindle mares from Texas in an open stock trailer and stopped to refuel at various gas stations, people would come up and peer in the trailer and

exclaim they had never seen anything like it.

Some Brindles have been sold without papers because they were not considered a "correct" color in their breed. This applied not only to Brindles, but to "palomino" Thoroughbreds and "crop-out" Paint and Quarter horses, etc. Fortunately the climate is much more favorable for unusually colored horses today.

Brindle has occurred in such diverse breeds as Arabians, Thoroughbreds, mustangs, Quarter Horses, Tennessee Walking Horses, German and Bavarian Warmbloods, Russian horses, Spanish horses, and also in donkeys and mules.

While dogs are known to have partial brindle patterns, it's not known if the partial striping patterns seen in some equines are the result of partial brindling (as with dogs) or just coat developmental variations, and are thus not inheritable.

Many people confuse the brindle pattern with dun factor markings (stripe down the back, barring on legs, and occasional, regular-spaced striping down the ribs). At one time, it was thought the brindle pattern was just a variation of dun factoring. Indeed, there have been many examples of horses that were probably carrying both dun factoring and brindle coloring. However, as can be seen from pictures of numerous Brindles, many do not have any dun factor markings whatsoever, indicating the two patterns are probably distinct genetically.

Brindle horses also have texturing in their coat similar to that seen in some Appaloosa horses. Sometimes the brindle pattern seems to be inheritable, especially in terms of coat texturing, but the expression of the darker or more intense pigment to make the pattern visible is highly variable, and even varies with individual horses seasonally and /or yearly. Sometimes the pattern seems to be composed of dark hair (black or brown), sometimes of white hair (roan or gray).

Very little is known about the genetics of the brindle pattern in equines at this time. Previously, it was thought to be a random mutation or coat developmental variation that was probably not inheritable. However, examples of Brindles have been found dating back over 100 years. Also, there are known Brindles that have reproduced the pattern, especially in terms of coat texturing.

Information collected since 1990 on Brindle horses is now shedding some light on the brindle pattern. It now appears there may be two ways in which a Brindle phenotype (outward appearance) can occur. In some horses, the pattern has not been inheritable, pointing to a possible mosaic or "Chimeric" origin (mixed genotypes in the coat from various causes), as seen in tortoiseshell cats. In other horses, the pattern has been shown to be inheritable. However, there could be several genes involved producing similar patterns (much as pinto/paint spotting can result from several different genes).

Indeterminate patterns are those found in horses, donkeys, or mules with some sort of streaking or texturing which resembles and could be the result of brindling, but could also be the result of some other pattern. For example, it isn't known if some streaking seen on the ribs of a dun horse is actually a result of brindle plus dun, or is just the result of extensive dun factor markings. Just as some streaking on a brown horse could be the result of brindling, it could also have resulted from dappling breaking up countershading. Another example would be of an animal supposedly who has brindling, but photos do not show it clearly, or it was photographed at a time of year (usually spring) when even normal colored animals may show variations in hair as they shed their winter coats.

The first record of the brindle pattern in horses seems to be by J. A. Lusis, in the publication "Genetica, Volume 23", 1942. In the article on "Striping Patterns in Domestic Horses", he details a brindle Russian cab horse from around the 1800s that was preserved and put in a museum. Reports of brindle or brindle dun patterns from the 1860s to 1870s in the Criollo horses of South America have been documented by writers such as Marrero, Pereyra, Solanet, and Odriozola.

The International Buckskin Horse Association information booklet (copyright 1977) came up with the classification it called "brindle dun" to describe horses with striping all over the body. It describes it as "A different and unique body coloration with stripes appearing over the barrel of the body with most, if not all, of the dun factor characteristics. Brindle duns show up in the Netherlands and (is) referred to as an ancient dun color. The peculiar body markings can appear in the form of tear drops or zebra stripes".

"Brindle dun" has now existed since 1971 as the description of a horse with striping all over the body. However, by going back and looking at the earliest reports of what is currently called the "brindle dun" color by the International Buckskin Horse Association, one can see that while brindle has often been associated with dun since the first reports of this rare and unusual color, it may turn out that there are actually two separate components. There is a wide variation in expression of dun factor markings, and some horses considered to be "brindle duns" could be the maximum expression of dun factor markings.

The "very highly marked" dun factor horses could also have resulted from a combination of "brindle" and "dun factor". Finally, some of the "brindle duns" are examples of the "brindle" pattern without any accompanying dun factor markings whatsoever.

Thus it now appears, that "brindle dun" is actually composed of two components – "brindle" and "dun", and the combination of both factors is what produces the classic "brindle dun" horse. However, because a "heavily marked dun factor" horse would phenotypically resemble a "brindle + dun" horse, it will probably always be difficult to distinguish between the two.

Brindle Progress: In 1988, Mary Jagow of Silver Cliff, Colorado, began organizing the International Striped Horse Association to collect information on various striping patterns. She had noticed four basic types of striping: Dun Factor Striping, Bay Striping (renamed "countershading" striping), Roan Striping, and Brindle. This organization is no longer in existence.

In January 1997, a web site was established for Brindle horses by J. Sharon Batteate to provide information on the rare equine brindle coat color, locate other brindle colored horses, and assemble information on them.

In 1998, Anita Garza began a new registry, the "Brindle and Striped Equine International". It accepts horses, ponies, donkeys, zebras, and their hybrids with the brindle pattern. The registry also accepts those with heavy and /or unique dun factor markings and animals with the "netting" pattern.

Vet Gen, a canine and equine genetic detection service company from Michigan, discovered in 2002 the gene that causes brindle in dogs. After that discovery they began collecting DNA samples of Brindle horses to see if there might be a gene that is roughly the same as the one in dogs, causing the brindle pattern in horses as well.

In the February 2006 edition of the American Quarter Horse Journal there is an article on Brindles titled "One in a Million" by Christine Hamilton, stating that they have finally verified two Quarter Horse Brindles as being of a Chimera origin. The testing was done at U.C. Davis in California. One of the horses was Denise Charpilloz's mare and the other was "Dunbar's Gold". J. Sharon Batteate is scheduled to meet with Cecilia Penedo at U.C. Davis for the purpose of Cecilia documenting Sharon's records on the two inheritable lines of Brindle she has worked with.

It would seem that the brindle pattern is not just some random oddity or abnormality. However, before starting to draw too many conclusions about the pattern, there is a need to locate more examples for study. There needs to be more work and study on brindle coloration.

"I want to try to breed for more Brindles, and perhaps restore this ancient coat pattern to the equine world as a viable choice, just like duns or paints are," says Denise Charpilloz. "I believe this is a potentially important contribution for all horse enthusiasts. I hope others are able to join me in pursuing the Brindle coat pattern."

Buckskin

The color of a Buckskin horse is an indication of the superior genetic heritage it possesses. Although it is a color, Buckskin and also grulla and dun horses are noted for many qualities that are not characteristic of other types of horses. The true Buckskin may trace his lineage through a direct line of dun or buckskin colored ancestors as far back as recorded history of the animal is available.

Buckskin is a clear golden color with a black mane and tail. It can be various shades of dark gold to yellow with dark brown or black points. It sometimes has a dorsal stripe down the spine from the mane to the tail. Buckskins breed true to their own color a large percentage of the time regardless of the color of their mate.

Having a common color, among other traits, is proof of a distinct pure breed. For instance, the Tarpan breed, one of the most unusual in the world today, was a prehistoric wild horse type that became extinct in 1876. Later genetically recreated in 1933, the breed was all mouse dun or grulla in color. Another example is the Polish Koniks, a breed preserved today by the Polish government. These small horses are grulla colored and identical. They reproduced to be of the same height, confor-mation and color so that they are very hard to identify individually as they all look the same.

The Buckskin is definitely a "breed" type and proof of this may even be found by viewing his hair under a microscope. When examining the color pigment in the hair shafts of both Buckskin and dun, the pigment deposits on both are arranged very much the same with the exception that the Buckskin has no concentration of pigment at the tip ends of the hair.

But the Buckskin horse is not a mere color, contrary to the belief of many. It is considered more than a color in the equine world. It has been long noted for its superior qualities and strength. It has more stamina, more determination, harder feet, better bone, and are generally hardier than other horses. A Buckskin with a weak or spavined leg is a rarity. "Tough as wet leather" is a good description of the true Buckskin.

The progenitor of the Buckskin was the old original breed of Spain, the Sorraia. The Norwegian Dun found today in Norway and other Scandinavian countries is also a probable contributor. (This breed is so old that its actual origin is lost in antiquity. However, there are many indications that even they obtained their dun coloring from the horses of

Spain.) The blood of the Sorraia (and the Norwegian Dun as well) filtered into nearly every breed in the world; hence the fact that the Buckskin, dun or grulla may be found in nearly all breeds today. Thus it is believed that the Spanish Sorraia and the Norwegian Dun are the breeds originating the Buckskin and dun horses of today's light horse industry.

The Spaniards brought the original Buckskin and dun horses to America when it was being settled. These horses played an important roll in the making of America and are a part of its heritage.

The registry for Buckskins, the International Buckskin Horse Association (IBHA), also includes dun, red dun and grulla body colors because all these colors trace back to common ancient ancestors.

Dr. Ben K. Green, an author of horse books, began researching equine colors as a veterinarian in Texas. He worked with IBHA to help develop the understanding and guidelines for registering Buckskin horses. His book, "The Color of Horses", was the first definitive book on horse colors. His research of 30 years began in West Texas and took him to South America, Europe and the Middle East. His interest in Buckskins started from the time he was a youngster when he heard old cowboys brag especially about the many good qualities of dun and buckskin colored horses. Dr. Green states in his book that dun, Buckskin and grulla horses have long been described by early day cowboys as being the toughest horses the West ever had.

The Buckskin and grulla (or grullo) colors are both types of dun. The dun descends from the pure breeds of Europe such as the Norwegian Fjord Horse, the Icelandic Horse, the Norwegian Dun and the Sorraia. The grullo (pronounced GREW-YAH) is the Spanish name for the sandhill crane, a slate colored bird. It would be safe to assume that the colors as

Grulla

they are known today came from the Spanish horses, if for no other reason than acknowledging where the names of the colors originated. The dun terminology is from the "Gateado" name in Spanish and the "Lobo Dun" from the Spanish word "lobo". The word buckskin is assumed to come from an early settler term describing the color of tanned deer hide.

Dun is a color that was always found among horses in the wild. Early wild horses of Northwestern America had many dun animals among them. Dun is rarely found in horses bred for racing and showing (i.e. Thoroughbreds and American Saddlebreds) except in Quarter Horses which, of course, derive their background from native-bred horses. Duns, like Buckskins, are known to have tough hides and durable feet.

The primitive markings found on horses today are referred to by the International Buckskin Horse Association as "Dun Factor" points. The terminology "Dun Factor" was readily adopted by other associations in the United States and Australia to refer to the markings. Those markings include the dorsal stripe ("list" or "eel" stripe) down the back of a horse, stripes over the withers and stripes across the knees and hocks. The stripes across the knees and hocks are sometimes called "zebra stripes". Very rarely a horse will be extensively striped, almost to the extent of a zebra, and those in the dun color are referred to as brindle dun by IBHA. Such stripped horses are known to occur in Siberia, Scandinavia and Argentina. There are only 4 or 5 such horses registered in the IBHA in this country.

Through natural reproduction in the wild, horses were found to produce a variance of shades within the colors. In the United States when wild horses and domesticated horses ran together with no human interference or control of breeding, dun color variances also occurred.

With the addition of the human element entering the modern day breeding programs, horses of desired pedigrees, show records, conformation, etc. are being bred for certain preferences and body colors and are mixed to attain the other preferences. Color is bred for and maintained more truly by those breeders for their choice of color breed registries. Examples are breeders within the registries of Paint, Pinto, Palomino, Buckskin, Appaloosa, etc. that are breeding not only for improved conformation and performance ability, but to maintain the color they like. Breeders choosing only to breed for pedigrees, conformation or performance ability with no regard for color sometimes create undesirable color or even undistinguishable color where new terminology is needed.

A gray-dun can be called just that, although the horse at a later age would be considered just gray. There is no name for a red dun going gray, but such a horse could appear to be a rose gray in later life. Yellow duns having the gray gene might look neither gray nor rose gray and would create some problem in description as they age. Some horses change color from season to season. However those horses changing in body shade with the seasons usually remain within the same color group. Some Buckskins will appear lighter or darker in winter months. Some grullas turn almost black in the winter months, but return to their body shade with spring shedding, maintaining the same color from year to year.

Regardless of the variance in shades, Buckskins, duns and grullas can do almost everything expected of Spanish descended horses: pleasure ride, trail ride, show in hunt seat or western performance, jumping and dressage are but a few. Cutting cattle, working on ranches, competing in gyhmkana, pulling a cart or buggy and used in rodeo events (roping, pickup riders, etc.) are other areas of competition, work, and enjoyment where they can be found.

The original concept of the International Registry was formed as the International Buckskin Horse Registry in California. It subsequently moved to Indiana in 1970 where it was incorporated the following year as The International Buckskin Horse Association. It registers buckskin, dun, red dun, and grulla horses. It has proven to be the largest and most progressive registry in the world for these horses. In 2002, IBHA entered into an alliance with the American Quarter Horse Association which provided additional exposure for Buckskins and duns and positioned IBHA as the acknowledged authority for those horses. Because of this, marketability of IBHA registered horses has increased.

The horse industry as a whole continues to grow. A great many people are seeking good breeding stock to add to their herds and they insist on registered horses. More people are looking for double registered horses and many IBHA horses qualify for double registration. Competition has grown in IBHA events and it is the way to go with the Buckskin, dun or grulla horses. IBHA registered horses have increased in value since the need was created to have such a registry.

IBHA has three registration classifications: Appendix, Tentative, and Permanent. The Appendix registration is for all foals.

The Tentative stage is the "proving ground" for stallions and mares. All stallions and mares over one year of age are registered as Tentative unless both parents are already Permanent. Tentative stallions must sire twelve IBHA registered foals and Tentative mares must produce three IBHA registered foals to be eligible for advancement to the Permanent stage of registration.

Geldings are eligible for direct Permanent registration. Stallions and mares are advanced to Permanent status when they have met the requirements.

COLOR CLASSIFICATIONS:
Buckskin:

A true colored Buckskin should be the color of tanned deer hide with black points. Shades may vary from yellow to dark gold. Points (mane, tail, legs) can be dark brown or black. Buckskin is a self color (solid color on the main body of the horse) and is clear of any smuttiness that may appear in horses of other body colors. Guard hairs which are buckskin colored grow through the body coat up over the base of the mane and tail. A Buckskin may or may not have a dorsal stripe. Dappling on the body color is acceptable for registration.
Dun:

Dun is an intense color with a hide that has an abundance of pigment in the hairs. Dun differs from the Buckskin in the respect that the body color is a duller shade and often will have a smutty appearance. Dun horses commonly have dark points of brown to black. Rarely a dun will be classified with lighter points or an admixture of light and dark hairs within the points. They sport the "dun factor" points which include dorsal stripe, shoulder stripe, and often leg barring will occur as well. Duns vary in body shades.

Grulla:

Grulla is also considered an intense color. The body color is described as mouse, blue, dove, or slate colored – with dark sepia to black points. Grulla has no white hairs mixed in the body hairs. The hide of a grulla is comparable to the hide of the dun and is well pigmented to withstand heat and sunlight. Grullas have the dorsal stripe and in most cases have shoulder stripes and leg barring. They are considered one of the most predominant species of the "dun factor". Grulla should not be confused with roan or gray colors.

Red Dun:

The red dun will vary in body shades of red in the range of peach to copper to rich red. In all shades, the accompanying points will be a darker red or chestnut and be in contrast to a lighter body color. Red dun must have a definite dorsal stripe to be eligible. The dorsal stripe will usually be dark red and predominant. Leg barring and shoulder stripes are common. Horses with faint dorsal stripes that do not appear on photos may be denied registration. The red dun or copper dun is categorized as a self color.

Brindle Dun:

Brindle duns have different and unique body coloration with stripes appearing over the barrel of the body. They have most, if not all, of the dun factor characteristics. Brindle duns show up in the Netherlands and are referred to as an ancient dun color. The peculiar body markings can appear in the form of tear drops or zebra stripes.

White Markings:

White markings on the face and lower legs are permissible. Horses having white markings on the body, other than face or lower legs that do not reflect Paint characteristics, will be considered individually. Blue eyes are permissible providing the body color and conformation is acceptable.

CONFORMATION:

The conformation of eligible horses can vary from the Arabian type to the "bulldog" Quarter Horse type. The ideal type is that of a western or stock saddle horse. There is no preferred type of conformation by IBHA. Owners may breed the type of saddle horse they enjoy most. A horse should be a good representative of the type it represents. Inferior quality or horses with undesirable inherent characteristics are not accept-

ed for registration. Horses showing predominance of draft horse blood are not eligible. Ponies are not eligible. The mature horse is to stand at least 14 hands tall.

Disqualifications:

Horses appearing to have some dun factor characteristics but not being of acceptable color cannot be registered. Chestnut, gray, bay, sorrel, or palomino with dorsal stripe will not be accepted as dun horses.

Any horses having albino, Appaloosa, Paint or Pinto horse characteristics are not eligible for IBHA registration. Palominos with dorsal stripes and line backed sorrels, chestnuts, grays and bays are not eligible. Matured horses under 14 hands are not eligible. Horses showing a predominance of gray hairs to become grayer and horses showing roan hairs throughout the body are not eligible for registration.

DUN FACTOR POINTS:

The Dun Factor points specified for a Dun Factor Class in a horse show shall be as follows:
1. Dorsal stripe
2. Leg barring
3. Ear tips and ears with edging
4. Shoulder stripe or shadow
5. Neck shadow
6. Cobwebbing
7. Face mask
8. Mane and/or tail frosting and guard hairs
9. Mottling

No one body color shall be preferred. Conformation is to be considered in judging as not less than 10% or more than 20%. Interfering white markings are not desirable, but are not prohibited. Interfering white is defined as white in an area that would interfere with a Dun Factor marking.

The dorsal stripe may be black, brown or red and will vary according to the body color. The stripe will run along the backbone from the withers to the base of the tail. Occasionally the dorsal stripe will not run the full length of the backbone. The width of the stripe will vary. The more pronounced, the better. A dorsal stripe with prongs or barbs extending from the sides is considered better than one without.

Leg barring is horizontal stripes of varying widths that appear across the hocks, on the inside and front of hind legs, on the back of forearms, and across the knees.

Ear tips and/or ear edges are when the color on the ends of ears is darker than the body color. Ears are usually outlined on the edges. The most pronounced of ears will have horizontal stripes on the backside.

A shoulder stripe or shadowing is a transverse stripe over the withers running down from the withers in varying widths and lengths. Occasionally more than one stripe is seen in different lengths. In some cases a large shadow effect is seen due to a large area covered or stripes close together forming the shadow.

Neck shadowing or neck stripes are usually dark areas through the neck extending into the hollow of the shoulder. Occasionally dark shadows will appear only on the crest of the neck or dark lines will point down from the base of the mane.

Cobwebbing originates on the forehead. Lines extend in varying lengths over the forehead resembling a spider web. Occasional lines extend from the eye in a misplaced "eyebrow" effect. Penciling may occur completely around the eye.

Face masking is black, brown, or red shading on the bridge of the nose with the same color usually around the eyes. The masking effect may spread to the jaw and muzzle or be outlined around the lips and nostrils.

Mane and/or tail frosting is light hairs on either side of the mane or interspersed throughout the mane. In the tail, light hairs appear at the dock of the tail and can run throughout the length of the tail. The frosted hairs may shed during summer months in which case they would reappear during the fall and winter. In some Buckskin horses the frosting will appear as white hairs mixed through the black mane and tail.

Mottling is not to be confused with dapples on a horse's body. Mottling is found on the forearm, gaskins, shoulders and stifles. It appears as a circular motif in shades darker than the body color. Mottling gives the appearance of "reversed" dapples. It is generally not found on the horse's winter coat.

Half Arabian and Anglo-Arabian

The Arabian horse has captivated people for centuries with its beauty and courageous heart. It was bred to thrive under harsh desert conditions and serve its master with unwavering loyalty. This endearing horse became a treasured and vital part of the Bedouin family and was cherished above all other possessions. From these early beginnings, the Arabian's personable nature and innate desire for human companionship have continued as hallmarks of the breed today.

As the oldest of all the light breeds and foundation stock of most, the Arabian is unique. The purity of its bloodlines enables it to complement any breed with which it is crossed. It carries bloodlines that trace back a thousand years, making it the purest breed of horse in the world. Arabian purity endows it with prepotence, or genetic strength, in passing on desirable traits to its offspring, such as stamina, refinement, intelligence and trainability. Consequently, horsemen through the ages used the Arabian as a fundamental building block in their breeding programs.

Half Arabian

So crossing purebred Arabians with other breeds is not a new concept. This is how all light horse breeds developed. The tradition of upgrading a breed through the inclusion of Arabian blood has produced some outstanding horses.

Today Arabian ancestry can be found in many recreational breeds. Lipizzaners, Thoroughbreds, Welsh Ponies, Appaloosas, Quarter Horses, Morgans, Saddlebreds, Tennessee Walking Horses and many others can trace their origins to the Arabian horse.

Half Arabian and Anglo-Arabian horses have also proven to be popular. There are 333,000 Half Arabians and 10,000 Anglo-Arabians registered with the Arabian Horse Association (AHA).

The Unique Half Arabian

The Half Arabian is bred to fulfill a variety of needs. Horses have increased beauty, endurance, willingness and intelligence when crossed with an Arabian. This gives riders unlimited choices for pleasure or performance. The multi-purpose Half Arabian is a refined, willing and able athlete. It can compete and win prize money at the local, regional and national level, which makes it highly marketable for breeders and owners.

Half Arabians are the result of one registered Arabian parent and a parent of another pure or mixed breed. Grade horses can also be bred to a registered Arabian and produce a registerable foal.

A wide variety of performance and pleasure Half Arabian horses have evolved. The combinations are as numerous as there are breeds of horses. For a classic saddle-seat equitation horse, the Arabian/Saddlebred is perfect. English enthusiasts also find exciting horses among the Hackney crosses as well as the Saddlebreds.

For dressage there is the expressive Arabian/warmblood. The growth of dressage, hunter/jumper and sport horse divisions makes the Arabian/warmbloods popular because they're lighter, elegant and more responsive. The addition of the Sport Horse Nationals to AHA national competitions provides a great showcase for many of these super athletes.

For a nimble reiner or cutter, the Arabian/Quarter Horse might be the ticket. Avid fans of reining, trail riding, cutting and working cow events like how Half Arabians can maneuver, respond, learn quickly and outlast other working western horses.

When the distinct beauty of the Arabian horse is mixed with one of the color breeds, it produces flashy horses such as the Arabian/Pinto, who turns heads wherever it goes. The Arabian adds eye-catching refinement to Buckskin, Palomino and Appaloosa horses as well as the Pinto.

The Arabian ranks as the top breed for endurance and other distance riding sports and many Half Arabians compete successfully as well.

Whatever cross is preferred, the Arabian can be counted on to add stamina, intelligence, beauty and a willing attitude. With more than 300,000 Half Arabians in North America, there is a horse with the look, way of going and personality to fit everyone.

HALF ARABIAN STANDARDS:

The Half Arabian cross is an effort to infuse the type, mentality and physical attributes of the Arabian with that of another breed. The purpose is to embody the positive characteristics of the Arabian as well as take on the positive attributes of the other breed.

Outcrosses to numerous different breeds have been successful. There is no stipulation as to what other breed an Arabian must be crossed with to produce a Half Arabian. The only criterion is that one parent must be a purebred Arabian.

When crossing the Arabian with a working type horse, such as a Quarter Horse or a Paint, the hope is to achieve a horse that has a strong working type body with a relaxed attitude for events such as Reining, western riding and Working Cow classes. However, it is important that these horses still retain the attractive Arabian look and trainable disposition. This is not to say that the outcross breeds do not have an attractive look or good disposition. It merely means that the desirable look of a Half Arabian horse is similar to a purebred Arabian but embodies the positive virtues of the outcross.

The same is true with Arabian/Saddlebreds or Dutch Harness horses for English type mounts as well as Arabian/warmbloods and Thoroughbred crosses for hunter or hack type horses.

A Half Arabian must have one registered purebred Arabian parent, either sire or dam, and the other parent may be a grade horse, a Half Arabian, or a horse registered with another breed. The purebred parent must be registered as a purebred Arabian with either the Arabian Horse Association (AHA) or the Canadian Arabian Horse Registry. A mule, hinny or any animal other than a horse is not eligible for registration. Horses conceived and born in the U.S. or Mexico can be registered with the AHA. There are no restrictions regarding height.

The AHA gives Half Arabian and Anglo-Arabian registration numbers from 1A to 9A depending on how much Arabian blood is present. Horses with one purebred Arabian parent and one non-Arabian parent would be considered Half Arabian (1A). Horses with one purebred Arabian parent and one parent that is Half Arabian are considered 3/4 Arabian (2A). Those from one purebred parent and a 3/4 purebred parent are considered 7/8's Arabian (3A), and so on up to 9A.

A horse that competes in Half Arabian classes should not be marked down for displaying Arabian characteristics, such as a high set tail carriage. The Half Arabian halter division is judged on conformation quality, substance, and Arabian type, in that order. The horse may show characteristics of any other breed, but the foregoing first named three qualities shall take precedence in adjudication of in-hand classes over breed type.

Half Arabian In-hand Division classes may be divided into "Stock/Hunter" and "Saddle/Pleasure" type. This makes it easier to compare "apples to apples" when judging. However, in classes that are not divided by discipline, no preference should be given to one outcross or the other. The afore mentioned judging criteria must always be observed.

The Anglo-Arabian Athlete

Picture the ultimate sport horse – substantial, strong, agile. Add the courage, heart and work ethic of the Arabian, and that is the jumper, eventer, endurance or dressage horse dreams are made of – the Anglo-Arabian. In combining two breeds – Arabian and Thoroughbred – the Anglo-Arabian acquires the intelligence, strength and aptitude for rigorous equestrian events. It has the mental focus and soundness of the Arabian with Thoroughbred size and speed. Two great breeds, one extraordinary sport horse!

The Arabian/Thoroughbred cross combination required for an Anglo-Arabian is created by one of the following: purebred Arabian and registered Thoroughbred, Anglo-Arabian and Anglo-Arabian, Anglo-Arabian and Thoroughbred or Anglo-Arabian and Arabian. An Anglo-Arabian may be a combination of Arabian and Thoroughbred blood with no less than 25 nor more than 75 percent of Arabian blood. Any more would qualify for registration as a Half Arabian as long as one parent was a purebred Arabian.

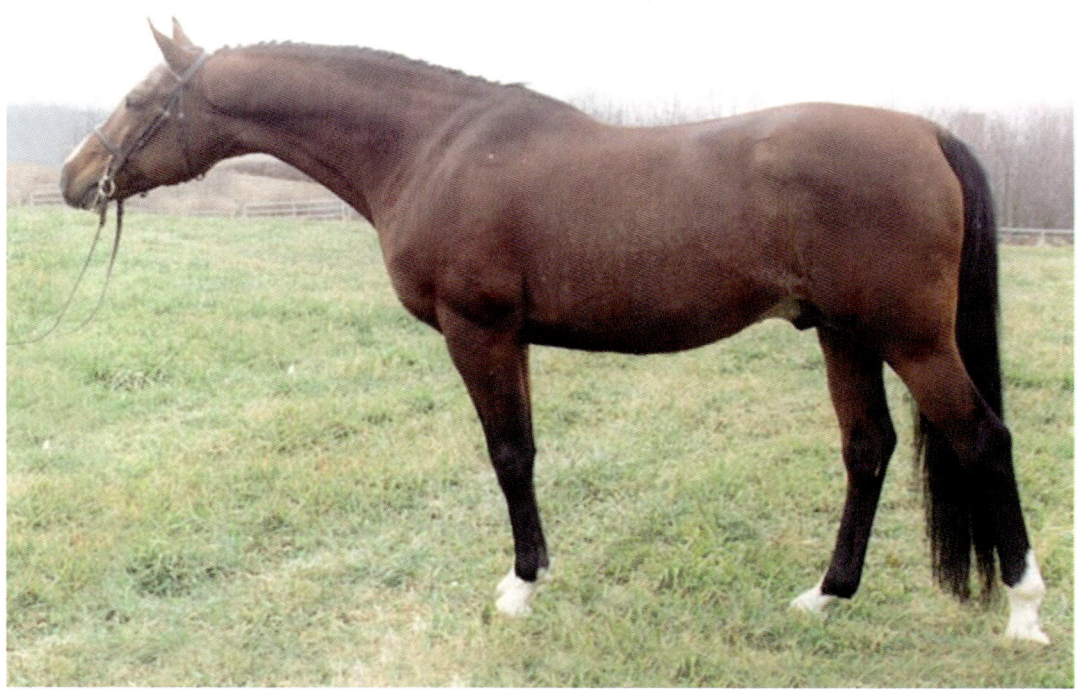

Anglo-Arabian

Anglo-Arabians in America: Early in the history of the Arabian in America, directors of the Arabian Horse Registry were sure that the best way to promote the breed was to get the U.S. Army interested. They spent a lot of time, money and energy proving that Arabians made the best cavalry horses by staging cross country endurance races. Arabians consistently won the races pitted against the cavalry's Thoroughbreds, which the U.S. had been utilizing. This convinced the Army of the Arabian's tremendous endurance ability and in the 1920s it added Arabian horses to its breeding program. Thus began the first significant introduction of the Arabian/Thoroughbred horse to North America.

Thereafter the Arabian/Thoroughbred cross continued to flourish, producing a superior breed known as the Anglo-Arabian. Serving as both transport and tank, the Anglo-Arabian had the speed and stamina to gallop endless miles across harsh territory and the courage to ride into battle.

Eventually the Army held the studbooks for the Anglo-Arabian. When the Army discontinued the cavalry after World War II, they sold the Half Arabian and Anglo-Arabian breed registries to what is now the Arabian Horse Association. The Anglo-Arabian grew from a few hundred horses in 1951 to 10,000 that are now registered.

Anglo-Arabians Today: The Anglo-Arabian possesses combined traits of two extremely athletic breeds. The Arabian horse adds stamina, intelligence and refinement while the Thoroughbred contributes its great racing ability and larger size. Because the breed is a cross between these two superb light horse breeds, it often expresses great courage and a willing attitude. It is a true aristocrat with the beauty of an athlete

The Anglo-Arabian offers a range of capabilities and each model embodies qualities that make it more than just a horse to ride. It has the speed, heart and stamina for Olympic sports and disciplines that require outstanding athletic ability. It makes a great sport or show horse in such competitions as eventing, jumping, dressage, fox hunting and distance riding. For the recreational rider who wants quality and a high fitness potential, an Anglo-Arabian is an excellent choice.

In England, the Anglo-Arabian is used for hunts, steeplechases and other sporting events. In Spain, riders appreciate the horse's courage in testing the stamina and fighting spirit of bulls destined for the ring. France dedicated its breeding of Anglo-Arabians to develope finely tuned sport horses that evolved into the Selle Francais. Many international and Olympic equestrian honors awarded to France for eventing, show jumping and dressage were won on famous Anglo-Arabian horses.

In the United States, the Anglo-Arabian has won top awards from national organizations in both endurance and dressage.

ANGLO-ARABIAN STANDARDS:

An Anglo-Arabian possesses distinct traits of both breeds. The Arabian intelligence, refinement and stamina mixes well with the Thoroughbred size and speed. The ultimate creation is an aristocratic athlete with competitive clout.

Recognized as a separate breed, Anglo-Arabians have their own registry through the Arabian Horse Association. They must have between 25 to 75 percent Arabian blood. To be eligible in the Anglo-Arabian Registry, the horse must be the result of one of the following crosses:

1. Thoroughbred stallion to purebred Arabian mare.
2. Purebred Arabian stallion to a Thoroughbred mare.
3. Anglo-Arabian stallion to Anglo-Arabian mare
4. Purebred Arabian stallion to Anglo-Arabian mare
5. Anglo-Arabian stallion to purebred Arabian mare
6. Thoroughbred stallion to Anglo-Arabian mare
7. Anglo-Arabian stallion to Thoroughbred mare.

Parents must be registered Arabian, Thoroughbred or Anglo-Arabian horses. The resulting foal must be no less than 25% Arabian and no more than 75% Arabian. If the foal is more than 75% Arabian, it is still eligible for registration within the Half Arabian Registry as long as one parent is a purebred Arabian.

The Anglo-Arabian halter division is judged on conformation quality, substance and Arabian type, in that order. The horse also may show characteristics of the Thoroughbred, but the first named three qualities shall take precedence in adjudication of in-hand classes over breed type.
<u>Average height:</u> 15.2 to 16.3 hands.
<u>Colors:</u> Any solid color is allowed, although several sabino patterned horses are registered.
<u>Disciplines:</u> Anglo-Arabians are eligible for competition in all Half Arabian classes. They excel at jumping, eventing, endurance, dressage and recreational riding.

As an Athlete:

A true adrenaline delivery machine, the Anglo-Arabian gives experienced riders the power on demand needed for strenuous equine sports. For competition in sport horse or endurance events or just as a durable horse with heart, an Anglo-Arabian is perfect. It's a high-octane horse who's always eager to see what's around the curve or over the next fence.

The Anglo-Arabian has the speed, heart and stamina for FEI disciplines that require superb athletic talent. Worldwide, it makes a perfect hunter/jumper, eventer, dressage or endurance horse. For hunter/jumper classes and eventing, the Anglo-Arabian has the size for long extensions, but remains easily maneuverable. It has substance and strength for clearing obstacles, but is neither slow nor bulky in its way of going.

It has the intelligence and temperament for the painstaking demands of dressage. The Anglo-Arabian projects a nimble elegance and commands attention with its charismatic presence. It captivates spectators and judges alike with its fluid movement and responsive attitude.

Despite its blue-blooded image, the Anglo-Arabian expresses its true form over long distances. Whether competing in a world-class endurance race over 100 mountainous miles or running 25 kilometers across the flatlands, the Anglo-Arabian is excellent.

For a performance potential in a recreational mount, the Anglo-Arabian delivers. His willing attitude gives active owners a quality horse for everyday riding.

Classy Competitors

Why do breeders create hybrid horses such as the Half Arabian or Anglo-Arabian? It is human nature to try and create something "new" and "different". Some might even call it improving on something, but others may say that it is not. Either way, no one can argue about the popularity and interest in Half-Arabians. Many Arabian enthusiasts that have purebreds also have Half Arabians and appreciate them both.

Most modern breeds of light horses are derivatives of a cross between an Arabian horse and another breed. These include the Thoroughbred and the American Quarter Horse. Were it not for innovative breeders, there would not be some of the outstanding breeds of horses enjoyed today.

Although many appreciate their Arabian horse crosses, whether registered or not, there are more recreational and competitive options with a registered horse and oftentimes it's easier to sell one that's already registered. A Half Arabian or Anglo-Arabian horse that is registered with another breed can be double registered. Thereby, Half Arabians and Anglo-Arabians retain increased value and marketability for resale.

The Arabian Horse Association continues to administer the Arabian, Anglo-Arabian and Half Arabian Registries. They offer more opportunities to compete, earn recognition and win prize money than any other partbred registry or organization. Horses compete in halter, performance, endurance and competitive trail riding.

There's a world of enjoyment for those involved with the Arabian, Half Arabian or Anglo-Arabian.

Morab

"Muscular, beautiful, proud and genuinely loving. It is easy to see why Morab owners treasure their horses and usually keep them for life. The refined, sculpted beauty of the Arabian, joined with the Morgan's dramatic, natural style and stamina, creates an elegantly powerful horse for use in the show ring, as a working horse, or on the trail. Indeed, most Morabs comfortably switch back and forth in all of these activities." Linda Konichek, one of the founders of the International Morab Registry, penned these remarks about the magnificent Morab.

Owners love the Morab disposition and personality; it is eager to please and easy to be around. Anyone who likes Arabians or Morgans also likes Morabs because it is a combination of these two treasured breeds.

The desire to unit the best traits of the Arabian and Morgan into one superlative horse has inspired breeders since the 1800s. Morabs express the beauty, spirit and endurance of the Arabian with the strength, power and common sense of the Morgan in a unique way. The fusion of these characteristics is genetically complimentary and gives the resulting Morab enhanced strength and depth from the Morgan with Arabian refinement and sensitivity. Muscular, yet refined, best describes the mature Morab.

Morab conformation consistently meets the ideal. Even newborn Morab foals are balanced and truly athletic, working naturally off their powerful hindquarters. In fact, a high percentage of equine veterinarians have chosen to own Morabs, citing the excellent conformation and over-all perfection of Morab legs and hooves, as well as the fact that they are very "easy keepers" and outstanding athletes, no matter what is asked of them. Foot and leg problems are a rarity in the breed and most owners choose to leave their Morabs unshod as their hard hooves require little maintenance; farriers are always impressed with the Morab hoof. Plus, Morabs combine high intelligence with a dependable, affectionate nature that is prized by all who own them.

The Morab is a graceful, free-flowing horse that is ideal for any type of riding. Many owners report that all they need to do between show classes is to change the gear to switch from English to Western or

Dressage; mature Morabs seem to welcome the change and stimulation to perform.

The Purebred Morab: Crossbreeding between breeds of horses can produce desirable high qualities, but often resulting foals, even from the same breeding, do not always show many like traits. This is not true of Morabs. Their qualities are based on foundation lines producing the purebred they are today, answering the question most often asked, which is their status as a breed. Well bred Morabs have proven over six generations that they transmit their genes with a high degree of certainty to their progeny. They display distinctive breed characteristics generation after generation, now obvious in fifth and sixth generation foals. In fact, a comparison of first and fifth generation foals of the same lines shows almost no changes; Morab breed characteristics remain strong.

One example of this is the Morab's inheritance of the Arabian skeleton, which is unique from other breeds in that it has 17 ribs (other horses have 18), 5 lumbar vertebra (other horses have 6), and 16 tail vertebra (other horses have 18). This skeleton finds its way into the Morab too, so there are two such breeds with this characteristic. It is this ability to transmit like characteristics to their progeny that makes the Morab a distinct breed, rather than just a very nice crossbred horse, putting the question of its breed status to rest.

HISTORY:

In the book, "History of the Arabian Horse Registry", written in the early 1900s, there is a provision for the get of the Arabian/Morgan crosses in the early Arabian Horse Club Registry.

In the 1857 book, "The Morgan Horse" by Morgan historian, D.C. Linsley, there is a lot more background. A major part of Linsley's essay was concerned with perpetuating and improving the Morgan breed, and stated where mares of Morgan blood could not be obtained, mares possessing a strain of racing or Arabian blood could be considered. Linsley specifically recommended 1/8 to 1/4 Arabian blood as suitable and many of the Morgan/Arab crosses (Morabs) could be found registered in the American Morgan Horse Association registry prior to their January 1, 1948 abolishment of Rule 2 that allowed outcrosses.

The Linsley essay has information about the first volume of the "Morgan Horse Registry" written by Colonel Joseph Battell. Battell's Volume I provided an entire chapter devoted to the stallion named "Golddust", a horse of great merit, whose bloodlines reveal he was a Morab, registered as #70 in the Morgan registry. Golddust was foaled in 1855 (bred by Andrew Hoke near Louisville, Kentucky) and sold when a weanling for $1000. His sire was "Vermont Morgan", his dam the unregistered "Hoke" mare. The Hoke mare was said to be by "Zicaaldi", a chestnut Arabian stallion, presented by the Sultan to the United States Consul, Mr. Rhind, who imported him.

Golddust became an important sire of the time. He was described as being pure gold in color, off hind ankle white, 16 hands high and weighing 1,275 pounds. It was reported he was never defeated in the show ring at the trot or at the flat-footed walk and that at the flat walk he could cover six miles in an hour. No stallion of his day produced larger, more handsome and showier horses or more winners in the show rings and trotting races of the era.

Golddust outperformed anything bred before in Kentucky. In 1861, Golddust raced and defeated "Iron Duke" in a match race, best three out of five heats, for a purse of $10,000. Besides being an animal of great beauty and refinement, he was noted for endowing his offspring with extreme speed. Although the Civil War and his own untimely death curtailed his stud career, he sired 302 foals and left 44 trotters of record. In "getting" speed (his progeny's racing records), he out ranks even the great "Hambletonian" (foundation sire to racing Standardbreds). In addition to their speed and racing quality, his get also illustrated the style and beauty of their Morgan and Arabian lineage. A search through the IMR™ (International Morab Registry) records has found over 100 of today's Morabs tracing their ancestors back to Golddust.

Little more was found until the 1920s, following World War I. At that time the famed publisher William Randolph Hearst had a superior Arabian breeding program and had a short-lived, but important Morgan breeding program, which included breeding Morabs. Hearst is credited with having coined the word "Morab" and some of his Morabs were registered as Morgans with the "Sunical" prefix (his Morgans had the "Piedmont" prefix), under the now extinct outcross Rule 2 of the American Morgan Horse Association (AMHA). Hearst bred Morabs through the 1930s and 1940s for work on the rough mountainous terrain of his ranches.

Hearst bred Morabs by crossing his mostly "Crabbet" or "Davenport" based Arabian stallions to his Morgan mares. To date there have been 25 of the IMR™ registered Morabs that trace back to this breeding program.

A significant early Morab breeding line was also developed by the Swenson brothers, Eric Pierson and Swen Albin Swenson, near Stamford, Texas on their father's (Swante M. Swenson) world famous SMS Ranch. The purchase of Morgan stud colts "Red Bird" and "Gotch", along with a band of seven Morgan brood mares just prior to 1920, marked the beginning of the SMS expansion into Morgans. A few years later, three Government Remount Arabian stallions were added to the stock and fine Morab cutting horses evolved.

Another origin of first generation Morabs centers on Martha Doyle Fuller of Clovis, California. In 1956, in an attempt to produce a horse that could successfully compete on the open show circuit, Mrs. Fuller developed a Morab breeding program. She had experimented with several horse breeds; however, the Morab was the only one she felt could consistently "fill the bill" generation after generation. It was from her program that the first Morab registry was formed.

Her son-in-law founded the American Morab Horse Association, Incorporated, on July 19, 1973. This first Morab registry was also called "Morab Horse Registry of America, Clovis" (after the town Martha lived in). In 1978 they established the 25/75 limit for Morgan/Arabian blood mix.

It was because of Mrs. Miller's efforts and this early registry that there are registered Morabs and a tie from the past to some of today's Morab breeding programs. The years since have brought a re-emergence and a new appreciation of the Morab breed for its own fine qualities.

Eventually from these registry foundations – Golddust, the Hearst breeding program, the SMS stock, and the Clovis registry – the International Morab Registry (IMR™) was created in 1992. The IMR™ is based only on Morgan and Arabian pedigree called Gold Seal, but it does accept any Morab that was registered with a prior Morab registry in an effort to protect breed history and maintain breed continuity.

Foundation stock ("first generation Morabs" from the early historical breedings) can be triple registered in the IMR Morab Registry, 1/2 Morgan registration in the Archival Morgan Record (AMR), and 1/2 Arabian registration in the Half Arabian registry of the Arabian Horse Association. This makes first generation Morabs not only triple registerable, but they have the added possibility of registration with various color and pattern registries. Their bloodlines are greatly treasured and will probably never die out. Succeeding generations from Morab to Morab breedings are no longer registerable with either half registry.

Other horses from Arabian and Morgan crosses can also be registered. Most registered Morabs today are from Morgan and Arabian parents, making them 50% Arabian and 50% Morgan. No Morab can go over 75% or under 25% of either Morgan or Arabian blood. They must have complete documentation of their Morgan and Arabian parentage. But there are many 3rd, 4th and even 5th generation Morabs from Morab to Morab breeding. Morabs foaled after January 1, 2003 must be DNA tested and Parent Verified. IMR™ registration certificates have a five generation pedigree, complete with dates of birth, color, and registration numbers of ancestors.

The IMR™ does not have any marking, color or pattern restrictions. Morabs can have a rainbow of colors and overo patterns (Pinto) as well as gaited movement. IMR™ is accepted by the Palomino Horse Breeders Of America and is pending with the Pinto Horse Association of America.

Sport, Combined Driving Endurance, show, endurance riding, dressage and eventing are now common to the breed. It has Partners 'N Performance award affiliations with the United States Dressage Federation, American Endurance Ride Conference, United States Combined Training Association and others are pending.

STANDARD:

The Morab body should be compact, medium length, well muscled, smooth and stylish. A Morab displays distinct refinement; the degree of refinement will vary with the breeding; however, it should always be apparent in the head and legs.

Morab refinement isn't the only inherited Arabian influence. The Arabian was often called the "drinker of winds" for its powerful lungs and endurance capacity. Combining its respiratory make-up with the broad, powerful chest of the Morgan provides the Morab with a naturally superior breathing system.

Morabs possess a shorter back than other breeds (one backbone less, as does the Arabian). This shorter back combined with the longer croup of the Morgan endows the Morab with great strength and a smooth graceful way of moving. They have a natural action or a lower action depending on the breeding of the animal. Morabs have a free-flowing gait, working off their powerful hindquarters with natural athletic ability and carrying themselves collected. This enables them to excel in competitive and endurance riding, as well as dressage, jumping and cutting. With their great disposition and way of moving, Morabs are very desirable as family and pleasure horses.

Morabs are late to mature, often not reaching their potential until the age of seven, but remain in top condition for many years.

Morab hooves and legs hold up well to excessive stress, since the Morgan contributes wonderful formation of bone and a medium-length pastern and the Arabian parent adds the broad, hard hoof and lower heel. This accounts for the almost non-existent foot and leg problems and many breeders report that their Morabs are never shod and require only minimum hoof trimming.

The Morab head may be straight to slightly dished with a big, powerful jaw in contrast to a small muzzle, with large, bold nostrils. The large dark eyes are set off by a wide forehead.

Morabs have a thick, luxuriant mane and tail which balances out their muscular build.

Despite all the other highly prized Morab traits, most owners and breeders will cite the Morab's intelligent, dependable and affectionate nature as its most valued quality. When the spirit and people-loving nature of the Arabian is added to the Morgan's, the resulting breed is a horse that cannot be beat in temperament, intelligence, and willing attitude. That is why mature Morabs are popular as mounts for children, amateur riders, and senior citizens.

QUALITY: The Morab should have dense bone with sufficient substance, well developed joints and tendons, and a fine, silky coat. Overall appearance of the Morab is always pleasing, showing great strength, but is never coarse.

DISPOSITION: Calm, affectionate, intelligent and dependable are the best descriptions of the Morab.

COLORS: Morabs may be any color and exhibit the white markings typical of a Morgan or Arabian: star, blaze, white stockings, etc.

HEIGHT: A mature Morab will generally range from 14.3 to 15.3 hands.

WEIGHT: 950 to 1200 pounds.

EYES: Morabs have large, dark, expressive, bright, clear and wide set eyes.

EARS: Set wide apart, characteristically curved in at the top, fine pointed and carried alertly.

MANE AND FORETOP: Full and silky in texture.

THROATLATCH: Clean and well defined, and never thick.

NECK: Heavy in appearance but refined and of good length, displaying a natural arched appearance. It should be smoothly joined to the shoulder and deepest at the point of the shoulder. Stallions tend to have a more fully developed crest than either a mare or gelding.

CHEST: Good depth and width. A mature Morab is broad in the chest which is quite noticeable through the heart, back ribs, and slightly wider through the hip. This room and compaction of body structure gives the Morab stamina as well as great speed.

WITHERS: Defined, not too high, but should be slightly higher than the point of the hip.

SHOULDERS: Muscled; of good length and slope.

BACK: Short, broad, deep in the girth.

FORELEGS: Long, sound, with flat bones and large joints, broad forearms, and short canon bones free of meat. Tendons should be squarely set and well apart; when viewed from the front, they should appear thin and must be straight; viewed from the side they will appear wide and strong.

FETLOCK: Large, not round, but rather wide.

BARREL: Large and round with well-sprung close ribs; deep and full, yet with a trim flank.

HIP: Muscled and of good length with a horizontal pelvic build that endows the Morab with a full, unsloped croup. The shape of the hindquarters and the pelvic angle is the most apparent difference between the Morab and other breeds. Hipbones never show on adults.

HIND LEGS: Squarely set and so placed that the Morab turns on its hindquarters with its legs well under it.

HOCKS: Neither close together nor wider apart than the fetlocks when viewed from the rear. Should be wide, deep and clean.

HOOF: Medium sized, nearly round, open at the heel, smooth and dense but never brittle.

PASTERNS: Clean, strong, medium length and should match the slope of the shoulder.

TAIL: Set fairly high, carried gaily.

WAY OF MOVING: The Morab in action has a free flowing gait, working off its strong hindquarters and carries itself collected. The Morab can possess a natural action or a lower action depending on the breeding of the animal.

Morgan

In 1789, George Washington became the first president of the United States and the U.S. Constitution was signed into law. That same year in Springfield, Massachusetts, a small, bay colt named "Figure" was born who was destined to play a dramatic role in the development of America. He would help build the new country and found America's original breed of horse, the Morgan. As American as apple pie, the Morgan was instrumental in forming the great nation of the United States and continues to be a great horse.

Figure's parentage was a mystery. Some thought he was sired by a Thoroughbred. Others claim the sire was a Dutch-bred stallion. While these details are lost to history, his style, beauty, intelligence and good sense guaranteed that he would not be forgotten. Given to a Vermont schoolteacher, Justin Morgan, as payment for a debt, he soon established a reputation for himself.

During his life, Figure's versatility as a working horse and prepotency as a sire earned him great respect in colonial New England. In keeping with the custom of the time, he also became known by his owner's name, Justin Morgan.

In the years to come, the accomplishments of the little bay stallion became legendary. His ability to work hard all day, move with agility over rocky fields and through dense woods, and still be fresh when the work was done, made him a popular stallion for breeding. His reputation only increased when he produced offspring in his image, displaying the same remarkable characteristics he had.

Like him, early Morgans earned a reputation for their strength and endurance. They soon became invaluable for clearing fields in New England and then beating all comers in trotting, running, pulling, and even walking races after a hard day's work. Some of Figure's offspring even held racing records.

Thus Figure founded the Morgan breed, America's first horse breed, and was a legend in his own time. From this one stallion and three of his get (the stallions "Woodbury", "Bulrush", and "Sherman") emerged a breed of horse that would secure a prominent place in American culture.

Used as light draft and stage coach horses in the 1800s, Morgans became known for their substance and stamina. As their fame spread, the breed moved westward, becoming popular as ranch and Pony Express horses where their intelligence and ability to work all day were valued.

When war divided the nation, the Morgan horse was the United States Cavalry's mount of choice. Vermont troops mounted on their Morgan horses were so envied that raids were staged to capture the horses for Confederate use. These Morgans were smaller than horses from other states like New York, yet they handled long marches and battles better than larger horses. Many times the troops and their horses would not have enough to eat, but the small, powerful Morgans could survive on just about anything. They remained level headed under fire and could march all day without becoming lame. Some books have even credited the Union victory to Morgans because they were quicker in battle than the large horses ridden by the Confederate troops of the south.

Union General Philip Sheridan rode his black Morgan "Rienzi" into history at the battle of Winchester, which was put to poetry by Thomas Buchanan Read's poem, "Sheridan's Ride." Confederate General Stonewall Jackson rode "Little Sorrel", a Morgan captured by Confederate troops and ridden by Jackson until he was killed in battle.

So well were Morgans suited to cavalry work that in 1907 the United States Government established an official Morgan breeding farm in Weybridge, Vermont. Morgans bred here were transferred to "remount" stallion stations across the nation where they were bred to local mares. This was to improve the quality of the offspring and ensure that quality horses would be available in times of war. These horses were used in both World Wars I and II.

When mounted units were being phased out in the early 1950s, the Government farm was deeded to the University of Vermont, where the Morgan breeding program continues. It is the oldest continuous breeding program of any breed of horse in the United States today.

The Morgan is the first recognized American horse breed. It is the official state animal of both Vermont and Massachusetts. Other breeds have claimed existence in colonial times, but today only Morgans can trace their bloodline to a common ancestor, while other breeds identify their foundation stock to horses alive at the turn of the 20th century, a full 100 years after the Morgan breed was born.

In the late 1800s, D.C. Linsley, a native of Middlebury, Vermont, researched the Morgan breed and compiled an essay on its history and genealogy. Using Linsley's work as a basis, Joseph Battell published the first volume of "The Morgan Horse Register" in 1894.

The Morgan Horse Club (now the American Morgan Horse Association) was organized at the 1909 Vermont State Fair (the fair was the national Morgan horse show of the early 20th century).

Today, 20 generations have passed, but the offspring of the phenomenal 18th century stallion, Justin Morgan, still share his remarkable traits. All Morgan horses trace their lineage to Justin Morgan, the only horse to have a breed named after him. This is unlike other American breeds which are based upon a foundation sire later in history or upon a group of horses chosen for similar traits such as gait, color or speed. The Morgan can trace its roots back to an earlier starting point than any other American breed.

Morgans have also distinguished themselves by making major contributions to the development of other breeds, including the American Saddlebred, Standardbred, Tennessee Walking Horse and Quarter Horse.

The Morgan's agility, stamina, beauty, and intelligence are inherited and valued in the American Quarter Horse. Popular Quarter Horse stallions such as "Joe Bailey", "Yellow Jacket", "Royal King" and "Joe Hancock" were half or full-blooded Morgans. Old timers present at the time can still talk of the train cars full of Morgan mares that were unloaded on the plains of the King Ranch in Texas to add Morgan traits to the ranch's cattle horses.

The Saddlebred world used an abundance of Morgan blood to develop their showy breed. Ninety percent of today's Saddlebreds still carry Morgan blood with its spirited attitude and flash needed to win in competition. The Saddlebred foundation stallion, "Peavine", was a grandson of the Morgan "Stockbridge Chief", and "Cabell's Lexington" was a grandson of "Morgan Blood's Black Hawk".

"Allen F-1", the foundation sire of the Tennessee Walking Horse, was out of the Morgan mare, "Maggie Marshall". Allen F-1 was bred to the mare "Gertrude", who is attributed to have Morgan sires in her pedigree, and produced the influential stallion "Roan Allen F-38". Today's Walking Horses still have the looks, endurance and personality the Morgan is known for.

The American Standardbred drew on Morgan sons and daughters to add stamina, substance and purity of gait to their trotting lines.

Each of these breeds has benefited from the Morgan's contributions of sound legs and feet, beauty, intelligence, and endurance. Not only did it contribute to other breeds of American horses, but the Morgan distinguished itself with strong, ground-covering gaits, making it perfect for traveling long distances in the New World. Its willingness to take on new challenges established its value with men who appreciated a good horse.

Today these same traits make the Morgan just as valuable to modern man. The Denver Police Department depends on Morgans to help patrol streets and control crowds. According to their officers, the Morgan will let himself be "talked into" difficult situations and has an above average intelligence. They also have "the best legs and feet of the entire horse world. They don't have the lumps and bumps like most patrol horses get, they just go, go, go!" One Denver officer said, "I believe that my patrol horse should go wherever I say to go, whether it's upstairs, downstairs, whatever. And the Morgan is that kind of horse. If I wanted him to swim the ocean, he would."

The Morgan is neither a fad nor a status symbol, but the sort of prized possession that makes even a first time owner become an undying promoter. His intelligence and good sense make him a perfect companion. His willingness and even temperament make him easy for all to enjoy; children or adults, individuals or families, amateurs or professionals. His soundness, athleticism and stamina make him a horse that gets the job done. His thriftiness and longevity have made this breed a bargain for more than 200 years. He is easy to love and affordable to own. With his proud carriage, upright and graceful neck, intelligent face, and kind eyes, he lifts the heart. He exists today because he has pleased his owners over the last two centuries. To please people is the Morgan's heritage.

He is a true aristocrat of the horse world!

IN THE SHOW RING: In bygone days, the Morgan earned his keep by clearing wooded mountainsides in Vermont. Yet after a full day of work, he could out walk, out trot and out run his challengers under saddle or in harness.

Today's show horses are now asked to specialize more often than they are asked to do multiple tasks, but Morgans can still be found winning in every arena of competition, from trail and jumping to reining and dressage. The versatile Morgan is a breed that can fill any role with the greatest success. Whether used for pleasure riding or competing at the local, national, or international level, the Morgan is the perfect horse. The animated excitement of the Park class, the mannerly way of the English Pleasure class, the smoothness of Pleasure Driving or Classic Pleasure riding, the ground-covering action of Hunter Pleasure riding, or easy going Western riding are all filled by Morgans with a winning style.

They excel in the roadster class and are unmatched in carriage driving. They have represented the United States in multi world-class competitions and have come home with numerous honors. Their speed, stamina and willingness to obey their driver in demanding situations makes them the most popular breed of carriage horse in the United States.

In the elegant world of dressage, Morgans have earned top honors against all breeds in national competitions. They are naturally balanced and can collect for precise movements with ease. Their medium size makes Morgans especially suitable for riders who want to enjoy all aspects of working with the dressage horse.

This same balance makes the Morgan uniquely qualified for the exciting world of reining. His agility and power produces winning sliding stops, spins and reining maneuvers. He excels at eventing and his stamina and endurance make him a champion in both combined training and competitive trail rigors. His ability and power make him an exciting mount when faced with the challenges of the jumper ring, where few can match his courage and intelligence. His compact size allows him to get in and out of tricky jumping combinations safely.

The Registry: In 1948, the American Morgan Horse Register closed its books in order to preserve Morgan type. It has established a reciprocal agreement from Morgans registered with the Canadian Morgan Horse Association and the British Morgan Horse Society. Today more than 160,000 Morgan horses have been recorded in the official register.

The Morgan is best known for its distinctive type which is still very much the same as that of Figure. Morgans can be distinguished from other breeds by their compact and muscular, yet refined bodies, their large expressive eyes, and their chiseled faces. Morgan "upheadedness", (a proud, upright head carriage), and stylish, spirited gaits are also recognizable traits of the breed.

In 1996, the Registry removed the High White Rule which restricted horses with white above the knees and hocks from being registered, as the consensus was that any horse that had two registered Morgans for parents should not be restricted from the record books. Morgans come in all colors and include black, brown, bay, chestnut, palomino, gray, creme and buckskin.

STANDARD

1. The head should be expressive with broad forehead, large prominent eyes and straight or slightly dished, short face; firm, fine lips; large nostrils and well-rounded jowls. The ears should be short and shapely, set rather wide apart and carried alertly. Mares may have a slightly longer ear.

2. The throatlatch is slightly deeper than other breeds but should be refined sufficiently to allow proper flexion at the poll and normal respiration.

3. The neck should come out on top of an extremely well-angulated shoulder with depth from top of withers to point of shoulder. It should be relatively fine in relation to sex. It should be slightly arched and should blend with the withers and back. The top line of the neck should be considerably longer than the bottom line. The stallion should have more crest than the mare or gelding. An animal gelded late in life may resemble the stallion more closely.

4. The withers should be well defined and extend into the back in proportion to the angulation of the shoulder.

5. The body should be compact with a short back, close coupling, broad loins, deep flank, well-sprung ribs and a long, well muscled croup. The tail is attached high and is carried gracefully and straight. A weak, low, or long back is a severe fault. The croup should not be higher than the withers.

6. The stifle should be placed well forward and low in the flank area.

7. The legs should be straight and sound with short cannons, flat bone, and an appearance of over-all substance with refinement. The forearm should be relatively long in proportion to the cannon. The pasterns should have sufficient length and angulation to provide a light, springy step.

8. The structure of the rear legs is of extreme importance to the selection of a long-lasting equine athlete. Any sign of poor angulation of the hocks, sickle hocks or cow hocks must be considered a severe fault. Lack of proper flexion of the hock is cause for very close examination of the entire structure of the rear legs and should not be tolerated in breeding stock or show ring winners.

9. The feet should be in proportion to the size of the horse; they should be round, open at the heel, with concave sole and hoof of dense structure.

10. Viewed from the front, the chest should be well developed. The front legs should be perpendicular to the ground and closely attached to the body.

11. Viewed from the side, the top line represents a gentle curve from the poll to the back, giving the impression of the neck sitting on top of the withers rather than in front of them, continuing to a short, straight back and a relatively level croup, rounding into a well muscled thigh. The tail should be attached high and carried well-arched. At maturity the croup should NOT be higher than the withers. The underline should be long and the body deep through the heart girth and flanks. The extreme angulation of the shoulder results in the arm being a little more vertical than in other breeds, placing the front legs slightly farther forward on the body. The front legs should be straight and perpendicular to the ground. The rear cannons should be perpendicular to the ground when points of hocks and buttocks are in the same vertical lines.

12. Viewed from the rear, the croup should be well rounded; thighs and gaskins are well-muscled. Legs should be straight. The gaskin should be relatively long in relation to the cannon. The Morgan should portray good spring of rib and well-rounded buttocks. Slab-sided individuals should be faulted.

13. The height ranges from 14.1 to 15.2 hands, with some individuals under or over.

14. Horses must be serviceably sound—i.e. must not show evidence of lameness, broken wind or complete loss of sight in either eye.

15. Stallions two years old and over must have all the fully-developed-physical characteristics of a stallion. Mature stallions must be masculine in appearance. Mares must be feminine in appearance.

16. Other distinctive attributes of the Morgan horse are his presence and personality. These include:

    1. Animation
    2. Stamina
    3. Vigor
    4. Alertness
    5. Adaptability
    6. Attitude
    7. Tractability

Correct way of going for an In-Hand class:

1. The walk should be rapid, flat-footed and elastic with a four-beat cadence and with the accent on flexion in the pastern.

2. The trot should be a two-beat, diagonal gait, animated, elastic, square and collected. The rear action should be in balance with the front.

3. Posing horses must stand squarely on all four feet with the front legs perpendicular to the ground. Rear legs may be placed slightly back. Judges must ask exhibitors to move the hind legs up under the horse for inspection.

Moriesian

The Moriesian horse is the result of a breeding program initiated in the United States to produce horses with the versatility of the Morgan and the elegance and charisma of the Friesian, two breeds from which it takes its name. The blending of these two magnificent breeds produces a combination that cannot be matched.

The power of the Friesian in conjunction with the manoeuverability of the Morgan creates an exceptional warmblood. Moriesians inherit their size and awesome presence from the Friesian and refined head and strong hindquarters from the Morgan. The classic beauty of both heritages is vividly displayed.

could carry large loads, exist on meager rations, and possessed the agility to be effective in battle. They were a heavy bodied, black, up-headed horse with an expressive face, high-set neck and outstanding crest.

The modern Friesian's distinctive trademarks continue in the breed with the high-set neck, outstanding crest, broad chest, lightly accentuated croup, low set tail and noble face. Equally impressive is the stunning, thick, flowing mane and long, luxuriant tail. Friesians also have feathering from the knees down. They have a rippling, heavier body mass with dense bone. The average height is between 15.2 and 16.3 hands. They are intelligent, graceful, agile, sweet

The Friesian horse reportedly dates back 3000 years, though the Freisian breed known today was developed in the 12th century in northern Europe. Friesians were ridden by the Teutonic Knights and used as war horses for the crusades. They were always a mount of the aristocrats, owned only by noblemen and knights. Their original breeding and pedigrees were very closely controlled.

Further refinement during the 17th century created a horse that natured and willing, with a powerful elastic gait. Whether used for jousting, dressage or driving, their beauty projects an awesome presence, leaving a lasting impression.

The Morgan horse is a distinctly American bred horse. More than 200 years ago a legendary horse breeder set out to create the ultimate utility horse. He began with the stallion "Figure", later known by his master's name, "Justin Morgan". The Morgan breed originated with this dark bay stallion born around 1789 in Springfield, Massachusetts. Justin Morgan was known for his strength, speed and stamina. He was as adept at clearing land and farm work as he was at challenging racing opponents in colonial New England.

Though little is known about his parentage, Justin Morgan is thought to have been of Thoroughbred extraction, but speculation includes a possible Arabian and Friesian heritage. It is rumored that there were several Friesian mares in the breeding barn at the time of his conception, which may explain why the Friesian horse blends so well with the Morgan horse.

Justin Morgan's ability to stamp his foals with his own traits was legendary. The Morgan breed grew and spread quickly as thrifty New Englanders recognized the same qualities in his offspring.

This one stallion, Justin Morgan, provided the basis for several of America's native breeds. The Morgan horse has been used in establishing such breeds as the Standardbred, American Saddlebred, American Quarter Horse, and the Tennessee Walking Horse.

The Morgan is easily recognized by its proud carriage, distinctive and attractive head with expressive eyes, and an upright, graceful neck that is muscular and crested. It is deep bodied and has compact, strongly muscled quarters and short, strong legs. The average height is 14.1 to 15.2 hands.

The Morgan breed exists greatly because of its people pleasing nature. It's also a flexible and maneuverable horse. Morgans are known for their versatility; they excel in park and pleasure riding or driving, dressage, jumping, trail, western riding, and cutting. Their stamina also makes them excellent in endurance riding.

The Moriesian, (pronounced "more ree jhin", similar to Polynesian) comes from these two exceptional breeds that are renowned for their trainability, mild manners and friendly dispositions. The Moriesian makes a perfect performance competitor, plus a quality family horse. The breed exhibits a kind and intelligent disposition. They are honest and willing to please. Their easy going temperament makes them a great companion for riders of all ages.

The Moriesian is also built for success in all disciplines with its refined, compact body that allows it to excel as a sporthorse. It displays a proud, up-headed stature, kind prominent eyes, expressive face, short back and dense bone structure. The slope of its shoulder and movement tends to be more Friesian-like, which gives it a regal appearance. People admire its luxurious full mane and tail, and some Moriesians exhibit the added bonus of "feathers" – long hair on their lower legs.

Although the breed is young, Moriesians typically excel with the athletic ability for winning both in harness and under saddle. They have the natural ability to collect into a frame of roundness for classical dressage and carriage driving. The combination of balance and symmetry they portray is truly art in motion!

Betty Pace, owner of Dark Knight Friesians in Ogden, Utah, started the Moriesian Horse Registry (MHR) in 1996. Betty saw the Morgan-Friesian cross as taking the best characteristics from each breed, resulting in a more versatile and lively horse.

Unfortunately, due to illness Betty was unable to maintain the registry and it became dormant for several months. In February 1999, while searching for a way to register a two year old horse, Barb Collins made contact with Betty, and eventually took over the registry. The spelling of the breed changed from Morisian to Moriesian to eliminate mispronunciations. The MHR then established rules and bylaws and contracted with the Veterinary Genetics Lab at the University of California, Davis, for DNA testing.

The registry soon developed a following. Ellie Neerdaels of Wisconsin produced "Opus Black Mikasa", an exceptional first generation registered Moriesian stud that has earned Elite Sport Horse status in the registry based on his USDF performance. He is also the first foundation stallion in the Moriesian Registry and the first to produce second generation Moriesians that qualify for the exclusive "M" division (registered Moriesian to registered Moriesian breeding). With his profound looks, disposition and talent, he turns many heads.

"My dream evolved to combine the incredible presence and qualities of the Friesian with the versatility of the Morgan, producing more refinement," says Ellie. "(This is done by) blending the best to the best to create size and substance along with refinement and the movement and balance I desired in a sport horse. With selective breeding, my horses turned out to be all of this – plus a pretty face! I get many comments from people who ask what breed of horse it is; many guess it is a Friesian, but were not sure, and are delighted to hear about this new breed. I remember long ago when the Morab breed was not taken seriously. With the interest I have seen that has been created from the versatility and disposition of the Moriesian, I see a promising future ahead for it."

Moriesians are known to continue to grow and mature until age six and it is not unusual for them to have a final change of height at that age. Their average size of 15.1 to 16.1 hands makes them comfortable for most riders.

Lighter boned than a Friesian, Moriesians have more of the qualities sought after in a sport horse. They have a full, upright, sloping neck set into an open shoulder with the capability of huge movement. They have the size and bone without being over-bulked for sport horse use; refinement and size without draft coarseness.

They make natural show animals that are magnificent to watch, ride and drive. They love to jump and many have shown themselves to excel in combined driving competitions and have proven apt competitors in classical dressage. They are versatility in motion!

Dave Wharton, representing Canada with his two Moriesians, competed in the World Pair Championship in Gladstone, New Jersey, in 1993. Mr. Wharton was one of the select few asked back for a special presentation before the judges. Also, Gloria Austin from Weirsdale, Florida, won ribbons with her four Moriesians in high level Four-In-Hand driving competitions.

As sport horses, Moriesians excel in dressage due to their presence and superb ability to naturally come under themselves and propel forward. The trot is up and forward with impulsion from the hindquarters. This comes naturally to Moriesians, as does collection and the ability to stretch and bend through the haunches, making dressage work seem easy. Their canter is big and forward and very comfortable. Lateral work seems easy for them.

Moriesians are adaptable and eager to perform. They make wonderful family horses. They love to learn, coming from two breeds known for their heart, mild manners, and friendly dispositions. Their common sense makes people think they are more mature at an early age. The breed is honest, willing to please and versatile, bringing lasting value to their owner.

The Moriesian Horse Registry has two main divisions and an Elite Status division:

1. **First Generation 'F' (Foundation) Moriesian Horse Registration**
   This division applies to Moriesians resulting from a Friesian and Morgan cross. DNA typing is required for all Moriesian mares and stallions.

2. **Moriesian 'M' Horse Registration**
   This division applies to the foal or horse from a Moriesian sire and Moriesian dam. Requirements include DNA proof of parentage.

3. **Elite Moriesian Sport Horse Division**
   The purpose of this division is to promote the Moriesian breed as sport horses. The horse must be a registered Moriesian and be proven in one of the following tests: In-hand, dressage, combined driving, show jumping (jumpers and hunters), endurance riding, competitive trail riding and performance.

Breed Standard:

1. The head should show a balance of Friesian nobility and Morgan refinement; the eyes are expressive and soft, the face is straight or slightly dished and short. The ears should be small and upright with turned in tips.
2. The neck should rise from a high set on the withers and show a clean arch to an upright headset. The throatlatch should be refined to allow flexion. The appearance is regal.
3. The withers should be defined yet blended into both neck and back cleanly.
4. The powerful shoulders should be a blend of depth from the Morgan with sloping angle of the Friesian. Movement to be free.
5. The back should be strong, well muscled, short to medium in length, and the barrel round and deep. A long or weak back is a fault. It should blend into broad loins and a well muscled and round croup. The croup should not be higher than the withers.
6. The legs should be straight and lighter boned than a Friesian, but more substantial than a Morgan with well-defined joints. The front legs should have forearms longer than the cannon bones and the rear legs have well muscled gaskins. Cannon bones are perpendicular to the ground. Legs may or may not be feathered, but feathering is preferred.
7. The feet should be size proportionate to leg bone mass. They are hard and strong.
8. The height of the Moriesian should fall between 15.1 and 16.1 hands, though variations are acceptable.

9. Any color is accepted, though most common colors are black, bay, and chestnut.
10. The horse's temperament should be kind, alert, and willing. Both heritages provide the Moriesian with a combination of stamina and versatility, calmness and loyalty.
11. All movement should be free and forward with suspension. The walk should have a distinct four-beat cadence and good length of stride. The trot should have a two beat cadence, be balanced, animated, and show engagement from behind and forward reach in front. The canter should have a three beat cadence, be well balanced and powerful with drive from behind.

Size and substance with refinement and presence! That is the Moriesian!

Palomino

The Palomino has journeyed down through the pages of history. There are stories of the "Golden Ones" linked to the crusades; the mail-clad crusaders saw them on the battlefield when they fought the desert chiefs of Saladin who rode them. There are stories about gold horses among the Arabs and the Moors. During the days of the crusades, the Emir Saladin, a Muslim leader, presented Richard The Lion-Hearted (Richard I from England) with two splendid war horses; one was a gray and the other a golden Palomino. There are other numerous leaders throughout history that have owned Palominos.

The place of origin of the Palomino probably never will be conclusively determined. Myths and legends of various countries shroud its beginning, which is no modern phenomenon. The gold horse with ivory-colored mane and tail appears in ancient tapestries and paintings of Europe and Asia as well as in Japanese and Chinese art of past centuries. Nowhere has the history of the Palomino been recorded, but most horsemen agree that all light bodied horses have descended from the Arabian and the Barb. There are many Arabian Palominos that are registered with the Palomino Horse Association and they are definite Palominos.

These splendid golden horses were favored by her Majesty Isabella de-Bourbon, that beloved queen who pawned her jewels so the expenses of the expedition which discovered the New World might be paid. In the Remuda Real (royal horse herd) of Spain, Queen Isabella kept a full hundred Palominos. As they were the chosen favorites of the crown, only members of the royal family and nobles of the household were permitted to ride them. A commoner might not even own one. It is on record that Queen Isabella sent a Palomino stallion and five mares to her Viceroy in New Spain, which is to say Mexico, to perpetuate the golden horse in the New World. From this nucleus, the blood spread to the Texas plains, and from Texas to California.

The word "Palomino" is a Spanish surname. Many feel that Palomino is only a color and not a breed, which is true that the color of Palomino comes in all breeds. But the Palomino of Spanish times, the "Golden Dorado", was as close to being a breed as any strain of horse. The Dorado was of Arabic-Moorish-Spanish blood and breeding, closely akin to the Arabian and the Moorish Barb.

The Palomino of Spanish times was not bred by crosses with sorrels. The Spanish had many shades of golden horses, and when they did use "corral breeding" (enclosing two horses for definite breeding purposes), a light color Palomino mare would be mated with a very dark-colored Palomino stallion. This point has been noted in an old book and printed in Barcelona in 1774.

In America, the Original Palomino Registry began in 1935 when Dick Halliday registered the golden stallion "El Rey de los Reyes". Mr. Halliday researched the golden horse for many years. He started writing magazine articles that brought the Palomino into public attention. His articles created a great deal of interest in the Palomino and within a few years hundreds of breeders were specializing in the production of horses with this color.

The Palomino Horse Association is the Original Palomino Registry incorporated in 1936. Today's Palomino Horse Association is the continuation of that registry. It originated in California, was moved to Missouri in 1972, and then to Pennsylvania in 1992. It has many sanctioned shows and members throughout the U.S., Canada and around the world. The registry includes horses from as many different countries. It doesn't discriminate against any breed. It recognizes all breeds based on color and conformation. There are horses from every breed registered with PHA. Also recognized are unregistered horses with the color that proves to be Palomino. The conformation would depend on the breed of horse, as all breeds differ in their size and other requirements.

The ideal color is that of a gold coin, but the shade can vary from light gold to medium or dark gold. The mane and tail should be white, ivory, or silver, but 15% dark or sorrel hair mixed in is allowed. In the last few years cream colored horses with blue eyes have been accepted. It has been researched and proven that these light colored Palominos always produce a Palomino. Therefore, they are definite breeding stock for the Palomino. Draft horses and ponies can't be registered, but are issued a Palomino draft or pony certificate under that heading.

White blazes and white stockings are allowed, but not spots. The body color must be solid gold. The gold color and flowing white mane and tail are outstanding and outshine most other colors. The mane and tail should be kept clean and cared for to maintain the white color. Bleaching them to make them white is not allowed.

The Palomino is a multi-purpose horse. They are admired not only for their beauty but for their versatility, maneuverability, and endurance. They are to be found in ranching, racing, rodeos, pleasure riding, parades, shows, fiestas, jumping, trail riding, and all other equine activities. There are even a few movie stars including "Mr. Ed", "Trigger", and "Trigger Jr.", which were registered with the Palomino Horse Association.

Mr. Ed starred in his own T.V. series. He was sired by "Chief Tonganoxie", a son of "The Harvester", a Palomino Saddlebred owned by Jim and Edna Fagen of California and used in many Tournament of Roses parades. Ed's dam was an Arabian named "Zetna" who was sired by "Antez", an Arabian imported from Poland. Ed was born in Puente, California.

Trigger and Trigger Jr. were the faithful and flashy mounts of the cowboy star, Roy Rodgers. It's not clear what Trigger's breed was, but Trigger Jr. was a Tennessee Walking Horse and both were registered with the Palomino Horse Association.

Many Palominos are ridden in parades where their beauty has brought them much popularity. The Palomino Association has had many horses that were ridden in the Tournament of Roses Parade and some still are today.

Patricia Rebuck, Vice President of the Palomino Horse Association, always had a love for horses, but especially appreciated the beauty of the Palomino. The Rebuck family have many trophies and awards won with their stunning Palominos. The flashy gold color and ivory manes and tails attracted attention wherever they went.

Pintabian

The Pintabian (pronounced pin-TAY-bee-an) is a horse that has certainly created quite a stir in the horse industry. Known for its timeless beauty, this elegant animal has distinguishing characteristics that set it apart from all other breeds in the equine world.

It was originally developed in North America by crossing horses with the tobiano (TOE-bee-AH-no) pattern back to Arabians until a strain of tobianos with Arabian "type" was created. These highly esteemed spotted horses are, in fact, over 99% Arabian in blood. It took dedicated horsemen with fortitude and foresight many years to bring this vivacious breed into existence.

and even included them as members of their households. These horses had their own "look". They were prized and guarded zealously by their ancient breeders who carefully protected the purity of the horses and painstakingly recorded their bloodlines.

However, purebred Arabians didn't have the tobiano spotting pattern that the Pintabian proudly sports today. Tobiano spotting was common among the wild horses of the vast Russian plains and many authorities believe tobiano horses were introduced there during the Roman Empire. Eventually horses with the same type of spotting could be found throughout Europe. Later, these particular markings were found among horses brought to North America and were associated with the American Indian who diligently sought them out. Native owners of such horses were regarded as affluent and materially wealthy.

The historic beauty of the Arabian and the flashy spotting of tobianos are combined in one exquisite breed, the Pintabian. The Pintabian Horse Registry, which was officially established in 1992, carries on the traditions of the early Bedouins by keeping accurate records and detailed

HISTORY

The Pintabian is indeed blessed with a colorful history.

The Arabian from which the Pintabian so strongly descends is considered the first breed of horse known to man. Bedouin tribes from the Middle East deserts believed the exquisite Arabian was a gift from God pedigrees. Its first sanctioned breed show was held in Detroit Lakes, Minnesota, in 2001 and drew exhibitors from as far away as Alberta, Canada to compete.

TOBIANO PATTERN

Tobiano is a specific and attractive, non-symmetrical pattern of large irregular markings. The well defined spots of a tobiano horse cover the body randomly, but white ordinarily crosses the topline at some point between the ears and the tail. The head is usually colored but often has the white markings common to those of non-spotted horses (such as a star, strip, blaze and /or snip, etc.) All four legs are generally white. The perfect example of this popular pattern is displayed by the Pintabian.

All tobianos are unique and can range from a largely white horse to that with very little white. Most breeders credit the ideal tobiano as being 50% colored and 50% white.

Because of the "simple dominant" genetics involved in this spotting pattern, at least one parent must be a tobiano to produce a horse with tobiano markings. A tobiano cannot be produced by two non-tobiano horses.

While tobiano markings are one of its distinguishing characteristics, the Pintabian is not a color breed based solely on that trait.

OVER 99% ARABIAN BLOOD

Virtually all modern-day light horse breeds are believed to descend with varying degrees from the Arabian. Experts in the field of genetics have known for years that it was possible to augment the captivating tobiano spotting pattern into the Arabian to produce the Pintabian breed.

It is important to note, however, that the Pintabian is not merely a Pinto/Arabian cross. As breeders across the United States and Canada will emphatically explain, these spotted equines do not have 50% Arabian blood running through their veins, as the uninitiated may incorrectly believe the name Pintabian indicates. Knowledgeable horsemen know that it took devoted breeders a minimum of seven consecutive generations of backcrossing horses with the dominant tobiano trait to purebred registered Arabians to achieve their beloved Pintabians.

Because the Arabian is considered one of the purest breeds and the Pintabian is over 99% Arabian, the Pintabian is one of the purest breeds in the world today.

The following two-step mathematical equation is used to determine the percentage of Arabian blood of any particular horse:
1) Add the percent of Arabian blood of the sire to the percent of Arabian blood of the dam.
2) Divide the result by two.

Crossbred example: If a Pinto stallion with no known Arabian blood was bred to a purebred Arabian mare, the result would be a crossbred that is 50% Arabian in blood.
(0% + 100% =100. 100 divided by 2 = 50% Arabian blood.)

Pintabian example: If a Pinto stallion that was verifiably 63/64th's Arabian in blood was bred to a purebred registered Arabian mare, the result, provided the offspring had tobiano markings, would be a Pintabian because it carries over 99% Arabian blood.

(98.4375% + 100% = 198.4375. 198.4375 divided by 2 = 99.21875% Arabian blood.)

The chart below shows the degree and percentage of Arabian blood of each backcrossed generation:

Generation		Degree		Percentage
1st Generation	=	1/2 Arabian	=	50% Arabian blood
2nd Generation	=	3/4 Arabian	=	75% Arabian blood
3rd Generation	=	7/8 Arabian	=	87.5 Arabian blood
4th Generation	=	15/16 Arabian	=	93.75% Arabian blood
5th Generation	=	31/32 Arabian	=	96.875% Arabian blood
6th Generation	=	63/64 Arabian	=	98.4375% Arabian blood
7th Generation	=	127/128 Arabian	=	99.21875% Arabian blood (Pintabian)

A minimum of seven generations of backcrossing tobianos to Arabians results in a relatively pure strain of spotted horses with a very distinctive appearance. These parti-colored horses have a standard set of characteristics, unlike crossbred horses which are hybrids that show the attributes of two separate breeds.

Pintabians consistently produce offspring of similar body type and disposition and, like their Arabian counterpart, are one of the purest breeds in existence.

STANDARD

Each Pintabian is uniquely marked and, quite predictably, physically resembles the Arabian. Traits include a head with a dished face and large, wide-set eyes. The neck is arched, the back is short and strong, the hip relatively level and the tail carriage high.

They generally stand between 14.2 and 15.2 hands and weigh between 900 and 1,100 pounds.

Pintabians come in a variety of colors which can include black, bay, buckskin, dun, chestnut, gray, grullo and palomino, along with, of course, the contrasting white.

Beneath their stylish external appearance is a horse that has the relentless "staying power" necessary for hard work. They are ideal for pleasure, showing, competitive and endurance sports, racing, driving, and so on. They make wonderful youth and family horses as well. They are known for their exceptional gracefulness, stamina, intelligence, versatility and good disposition. Their gentleness, tractability and willing attitude make them suitable for children as well as adults.

Horses that possess the desired tobiano markings and that are over 99% Arabian in blood (but less than 100% Arabian as purebred Arabians do not come in the tobiano pattern) are registered as Pintabians in the Colored Division of the Pintabian Horse Registry, Inc. At least one parent must be a tobiano to qualify for registration as a Pintabian. Horses that are less than 99% Arabian, but have spots, can't be registered as Pintabians.

Horses that are over 99% Arabian in blood (but less than 100% Arabian) without the desired tobiano markings may be used to produce Pintabians and are registered in the Breeding Stock Division. When horses registered in the Breeding Stock Division are bred to Pintabians registered in the Colored Division, the resulting offspring are over 99% Arabian in blood. These offspring, if tobiano marked, are registered as Pintabians in the Colored Division. If these offspring do not have tobiano markings, they may be used to produce Pintabians and are registered in the Breeding Stock Division.

Purebred Arabians may also be used as outcross horses to allow for additional bloodlines and are registered in the Arabian Outcross Division. These horses are of purebred registered Arabian heritage and are generally solid (without spots) but may possess patches of body white.

PINTABIAN BEAUTY

After much time and effort, foundation Pintabian breeders have finally seen their hopes and dreams come to fruition. They have successfully attained their goal of producing horses with the ability to perpetually pass on the many fine qualities associated with the Arabian, but with the added dazzle and appeal of spots, to future generations. In addition to North America, Pintabian horses can be found in Africa, Australia and Europe.

Their notable physical features combined with their expressive markings make it impossible to mistake the Pintabian horse for a member of any other breed.

In the world of horses, as with anything else, beauty is in the eye of the beholder. Cal Rector, President of the Pintabian Horse Registry, Inc., is quick to declare, "In my opinion, Pintabians are the most beautiful horses in the world today!"

Pinto Horse

The Pinto Horse is a refreshing change of pace from the average horse seen in a pasture. They are attractive because their colorful patterns are distinctive and unique. It's easy to pick out Pintos. Their spots can be seen from a distance, setting them apart from everything else. Even within the breed, their color patterns are different from each other. No two look exactly the same.

But they are unique in more ways than color. Pintos have diverse styles of types, enabling them to exhibit a range of versatile talents. Owners are pleased with the differences and the attention their beautiful horses attract. To have a horse that can never be mistaken for any other is a great feeling of pride and appreciation for a breed that stands out noticeably from all others.

HISTORY:

Studies of historical art reveal the early existence of what is recognized today as the Pinto Horse: a horse whose spotted coat pattern is comprised of white areas combined with another of the basic coat colors, making each Pinto unique.

Horses with Pinto markings appeared in ancient art throughout the Middle East. There is also evidence of tobiano patterns among the wild horses of the Russian Steppes, suggesting that the introduction of Pinto coloring spread to Europe possibly as early as during the Roman Empire. After that and for many years, European breeders crossed their native stock with Barb horses. European explorers, but chiefly Spanish explorers, brought their light riding stock including Pinto spotted horses to America. In time, some of their horses escaped and eventually spotted mustangs ran wild on the American range.

Meanwhile early Americans imported many of the well-established and stylish European breeds as foundation stock. Over time, however, with the civilization of the Native American and the white man's migration to the frontier, it became necessary to cross these fancy, but less suitable breedss of the Eastern seaboard with the wild mustang stock. This was to increase size and attractiveness as well as availability of horses better suited to the strenuous working conditions of the day. Eventually

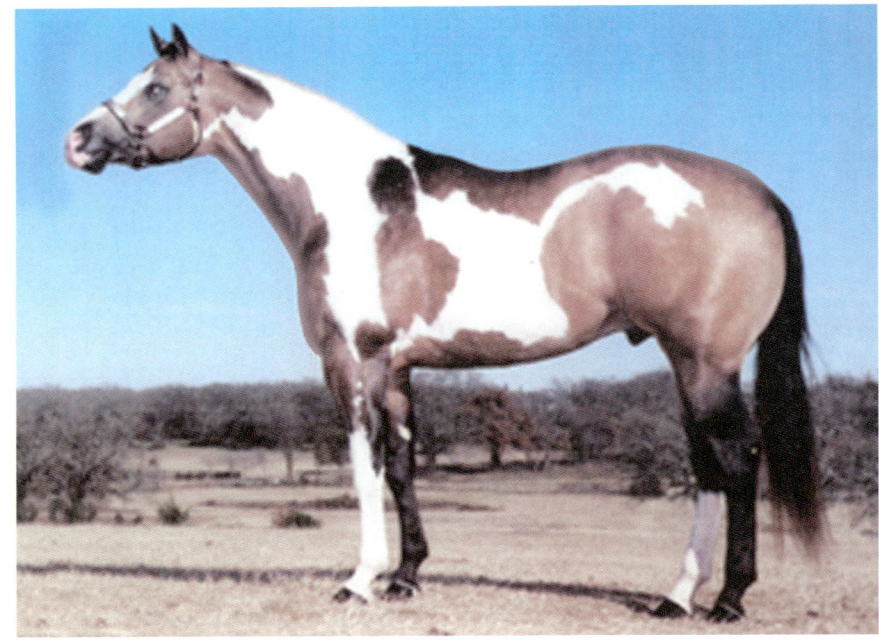

Overo Stock Type

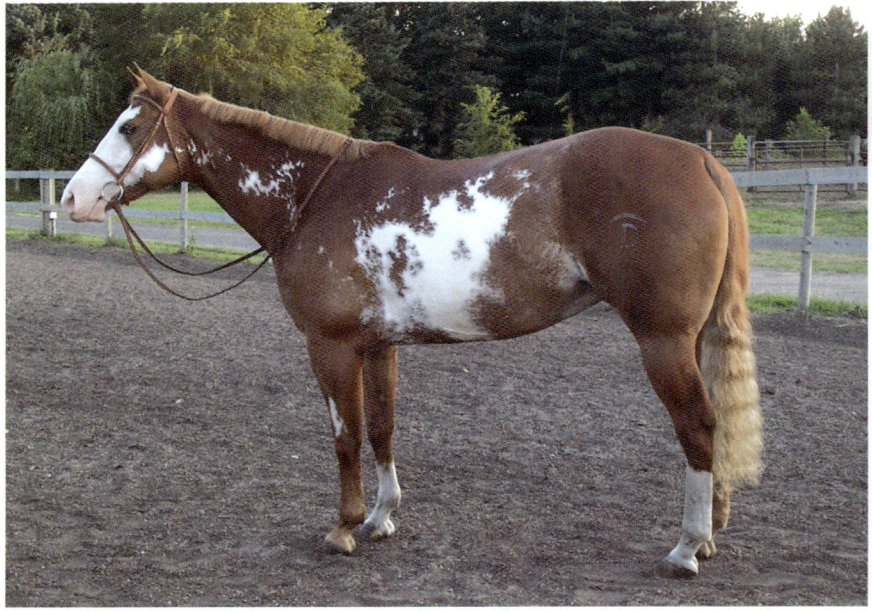

Overo Hunter Type

Tobiano Pleasure Type

licity to unit Pinto fanciers.

Some of the breeders associated with the Pinto Horse Society continued breeding with the principles set down by that organization, even after its demise. The Society had created a great deal of interest in the Pinto horse, particularly in the California area where it was based.

However, Pintos with cold-blooded ancestry continued to be bred in great numbers with poor quality. As a consequence, the general public and horse show judges associated the word "Pinto" with loudly marked but poorly conformed animals. By the early 1950s, this attitude had become so strong that even well built Pintos could not place in a show, or they had to work twice as hard to warrant even a glance from the judge.

This situation was covered very well in "Plight of the Pinto," an article by Kay Heikens which appeared in the December 1954 issue of The Western Horseman. Kay was speaking from personal experience of discrimination against the Pinto as well as citing several episodes involving other Pinto exhibitors. The article inspired others throughout the country to write of similar experiences.

These articles and letters aroused interest in starting a Pinto Horse association and registry. Eventually the Pinto Horse Association of America (PtHA) was established as a registry and incorporated in 1956 to encourage the promotion of "quality with color" in horses, ponies, and miniatures.

An announcement of readiness to accept registration applications for Pintos of all types was placed in The Western Horseman and Horse Lovers magazines. The response was astounding! More than 40 letters per day were received by PtHA.

Members decided to register Pintos with a wide variety of types so that suitable Pintos could be bred and shown in all areas of horsemanship. However, PtHA stressed the importance of fineness in the conformation of horses accepted for registration. This emphasis has contributed immeasurably to the improvement of the flashy horses throughout the country. Breeders hoped that the Pinto would be better accepted and that all their hard work in selective breeding would pay off.

Their efforts did succeed. Today PtHA maintains a registry of more than 125,500 horses, ponies, and miniatures throughout the United States, Canada, Europe, and Asia and the numbers continue to grow on

great wild herds infused with the flashy color patterns began to develop across America. They became associated with Native Americans, who domesticated and greatly valued them for their legendary magical qualities in battle.

Thus the Western-bred horse became a fixture of America, especially the uniquely marked Pinto whose colorful presence in parades and films always added a little extra glamour. Exhibitors crossed spotted horses with other breeds to produce specific types with good color.

In 1941 the Pinto Horse Society was formed, but it slipped into inactivity. Other attempts at registries for Pintos seemed to fail because they attempted to restrict conformation type or failed to develop enough puba daily basis. There are a variety of classes for an equally wide variety of Pintos. From halter classes that celebrate the conformational beauty of the Pinto to the vast array of driving and riding disciplines in performance classes, spectators can witness Pinto diversity. The combined purses from the 1998 International Futurity and Breeder's Cup Futurity alone approached $20,000.

Other classes are Dressage, Competitive Trail and Endurance Rides. With so many different types of Pintos represented in such a wide assortment of classes, it is clear to see the breed offers something for everyone.

Of the many questions posed to the PtHA, the most frequently asked is, what is the difference between Pintos and Paints? The big difference between the Pinto and Paint Horse (registered by the American Paint Horse Association) is the Paint Horse is limited to horses of documented and registered Paint, Quarter Horse, or Thoroughbred breeding. The difference in eligibility between the two registries has little to do with color or pattern; only bloodlines matter. Most Paints can be double registered as Stock or Hunter type Pintos, but PtHA also allows for the registration of miniature horses, ponies, and horses derived from other breed crosses, such as Arabian, Morgan, Saddlebred, and Tennessee Walking Horses, to name but a few.

BREED CHARACTERISTICS

The Pinto Horse is a color breed with documentation of pedigree. The color is that of spots that the horse is born with and that never change: in other words, Pinto spots aren't confused with Appaloosa spots, which can change as the horse matures. Certain restrictions and exclusions may apply, depending upon the sex, classification and background of each animal.

There are two recognized Pinto color patterns:

1) Tobiano (toe-bee-ah-no) appears to be white with large spots of color which often overlap and with a greater percentage of color than white. Spots of color typically originate from the head, chest, flank, and buttock, often including the tail. Legs are generally white, giving the appearance of a white horse with large or flowing spots of color. Generally, the white crosses the center of the back or topline of the horse. It is considered necessary to have a tobiano parent to achieve a tobiano foal.

2) Overo (o-vair-o) appears to be a colored horse with jagged white

Tobiano Saddle Type

markings usually originating on the animal's side or belly, spreading toward the neck, tail, legs, and back. The color appears to frame the white spots. Thus, an overo often has a dark tail, mane, legs, and backline. Bald or white faces often accompany the overo pattern. Some overos show white legs along with splashy white markings, seemingly comprised of round, lacy white spots. White almost never crosses the back or topline. A horse of Pinto coloration descended from two solid colored parents of another typically solid colored pure breed is called a "crop-out" and is of the overo pattern.

Color Requirements for Registration: The color requirement for horses is a total of at least 4 square inches of white hair with underlying pink skin on the body or on certain designated areas of the head (i.e. eligible zone). This is compared to Pinto ponies, which must have 3 square inches of white and miniatures must have 2 square inches. Animals with insufficient qualifying color to be accepted in the regular color division but with at least two or more Pinto characteristics, OR solid colored animals with documented color within two generations may be eligible for registration in the Breeding Stock Division.

Pedigree Requirements and Restrictions: While PtHA accepts animals derived from many different approved breed/registry crosses, it does not accept animals with Appaloosa, draft or mule breeding and/or characteristics.

Horse stallions must have both sire and dam registered with PtHA or another approved outcross registry.

Mares and geldings can be registered on their qualifying color alone.

Type Designation: Type is determined by the conformation and background of each horse/pony. All registered Pinto Horses and Ponies are identified within one of the following four types:

The Stock Type Pinto is an animal suitable for:
1. Western events
2. Hunter Seat events
3. A variety of other events.
The Stock Type Pinto should display the conformation associated with Quarter Horse breeding. Generally, double-registered Paints (APHA) will be registered in this division.

The Hunter Type Pinto is an animal suitable for:
1. Hunter Seat events
2. Western events
3. A variety of other events

The Hunter Type Pinto should display the conformation associated with Thoroughbred, approved Warmblood or running Quarter Horse breeding.

The Pleasure Type Pinto is an animal presented in a natural manner and suitable for:
1. General Western, English, and driving events
2. A variety of other events.
The Pleasure Type Pinto should display the conformation associated with Arabian or classic Morgan breeding.

The Saddle Type Pinto is an animal suitable for:
1. General English, Western and driving events
2. A variety of other events.
The Saddle Type Pinto should display the conformation associated with American Saddlebred, Tennessee Walking or Missouri Fox Trotting Horse.

There are four classifications of registration: One for horses, one for ponies, one for miniatures and one for B miniatures. Pinto Horses are defined as those that mature over 56 inches (14 hands) in height at the withers. (Only Pinto Horse classifications are listed in this book, not the ponies or miniatures.)

By providing two height divisions for miniature horses and four distinctly different conformation type divisions for horses and ponies, there truly is "Something for Everyone" in the Pinto. Each division, having its own rules and standards, allows for exhibition against "like" conformation and styles.

At some shows, Color Classes are offered. Tobianos and overos are judged on the most ideal markings of the individual Pinto. Ideal markings are defined as a 50-50 distribution of white and color overall on the Pinto. Classes specifically identified and offered as Color Classes for these patterns are not judged on conformation, as are other normal halter classes (in part).

In summary, the Pinto is the breed choice for equine enthusiasts of virtually all disciplines and events or just for pleasure riding, to get a job done, or when a horse, pony or miniature is to be used or shown with color and eye appeal.

Quarab

Straight Type

Pleasure Type

The classy grace of the Arabian horse and the substance of the Quarter Horse or Paint – that is the Quarab. For many years, breeders have been combining this unbeatable mix to produce one of the world's most amazing equines. They found that the Arabian's endurance and spirit uniquely blends with the body, usability and stability of the Qarter or Paint Horse, producing a superb all-around using horse.

Long time breeder, Ginny Hanks, states, "The breeding of this cross for over 40 years gives me the experience to say it is truly the only breed that can do it all."

Another long time breeder, Kristi Kraling, agrees. "The Quarab has the typical Arabian intelligence with the Quarter Horse calmness. It makes a really nice cross. They are extremely trainable, not spooky; they are smart, quiet and easy to handle." Some of her Quarabs were doing fig-

ure 8's in all three gaits when started under saddle on their 3rd ride ever. Her Quarab gelding earned the respect of a local Amish farmer as the best ever for pulling in harness or just riding.

The Quarab's popularity is expanding to new generations as more and more are introduced to and appreciate its outstanding characteristics. It is talented and versatile in every discipline; in the showring, on the trail or on a working ranch. The stamina from the Arabian and the muscling of the Quarter Horse or Paint enables it to excel in reining, cutting and long distance riding.

The Quarter Horse and Paint side allows breeding for colors like palomino, buckskin, grullo and Pinto, plus all the basic colors. So along with their versatility, Quarabs display interesting varieties of coat coloring.

The Quarab is also known for its loving and willing disposition. They are always eager to please and willing to keep trying. They are extremely smart, easy to handle and love to be with people.

REGISTRY

The breed has been around longer than its registry. American Quarter Horse Association records show that the Arabian stallion "Indraff" sired two half-Arabians (Quarabs) who were registered as Quarter Horses: "Indy Mac" and "Indy Sue" were both out of the Quarter Horse mare "Cotton Girl". Indy Sue was foaled in 1953 and earned 44 AQHA open performance points in six events, including Western Pleasure, Reining and Hunter Under Saddle. In 1960 she earned an open Performance Register of Merit. She had earnings with the National Cutting Horse Association as well. She was bred on to produce three registered Quarter Horse foals.

Indy Mac became a champion on the hunter circuit. There are also reports of a few purebred sabino Arabians that were inspected and registered in the American Paint Horse Association before their book was closed in the early 1980s.

The United Quarab Registry was the first registry and stud book specifically created to register the offspring of a purebred Arabian and a registered Quarter Horse. The UQR was the vision of Tamara Grant of Ogden, Utah, which she began in 1984. In 1989 the Painted Quarab Index was added to register those horses having a Paint Horse sire or dam rather than Quarter Horse. It was a privately owned studbook that was transferred and eventually it went out of business.

Then in 1999, a new association was formed, the International Quarab Horse Association (IQHA). It was started by Lisa Striegle with urging from several other owners and breeders and under the direction of the original guidelines set in place by Tamara Grant in 1984. The first year, the association was strictly a newsletter organization with a breeder's referral service, but a new studbook was created in 2000. The IQHA was incorporated in the state of Michigan in early 2003. Since then it has expanded to international proportions, including Alaska, Germany and the Netherlands.

An example of this new international Quarab interest is "Tishyno", a top IQHA/U.S.A. registered Quarab stallion in Germany. Having passed Germany's strict stud performance examination, he was registered in the First Studbook by ZfDP, the Breeding Association of German Horses. (Although Quarabs can only be registered in America, many foreign countries offer other registry options.) Being registered in Germany's ZfDP First Studbook means that a horse is judged to be a perfect stallion with good bone, hip, movement, etc. If a stallion is judged to be not as perfect, but still very good, he is registered in the Second Studbook. So the First Studbook holds the ultimate prestige for those who qualify for it, such as Tishyno.

Tishyno proves Quarab popularity with more than 80 offspring, including spotted foals seven years in a row and many premium foals. He won several distance races and classes in Western Riding and Open Reining. One great victory was in a 60 km distance race (37.28 miles) he did in 3 hours and 10 minutes. Being trained and ridden only by an amateur, he won Reining classes and beat top professionally handled German and European reining horses at the German Arabian Cup and at his first NRHA (National Reining Horse Association) show. Many avid horsemen at the competitions are impressed with the stallion. He is famous in Germany, being featured in various magazines, and has been establishing a foundation to the breed in Europe.

Tishyno typifies the wonderful diversity of the Quarab and the excitement it is creating as its influence continues to spread in Europe as well as North America.

BREED STANDARD

The Quarab is a horse that has solely Arabian and Quarter Horse (AQHA) or Paint Horse (APHA) blood. Crosses of 1/8 to 7/8 percentages in favor of either breed type are allowed. No other blood types are allowed.

Quarabs should display the conformation of a good saddle horse. They should appear well muscled, yet smooth and refined. They are compact and of medium length. Quarabs should exhibit a degree of refinement that will vary with their type – Straight, Stock or Pleasure. However refinement should always be present in the head and legs.

The overall appearance of the Quarab should be pleasing, never coarse, exhibiting good substance. The body should display good strength with sufficiently dense bone and well developed joints and tendons. The coat should be fine and smooth.

A mature Quarab will generally range from 14.2 to 16 hands,

although rare individuals may be slightly shorter or taller depending on the breeding. Weight is generally 900 to 1200 pounds.

Quarabs may be virtually any color and may exhibit the white markings common to Arabians or Quarter Horses. Recognized colors include: chestnut, black, bay, palomino, buckskin, cremello, perlino, smoky creme (double creme dilution on a black base), red dun, dun, grullo, sorrel, brown, and champagne (of various shades). Roans and greys are designated with their base color: bay grey, black grey, sorrel grey, etc.

Recognized patterns include roan, gray, tobiano, overo, and tovero. Roan and grey are uniquely listed as patterns. A Quarab may be described as sorrel grey tobiano – meaning "born sorrel, but turning grey with tobiano superimposed". This is especially helpful to those who are specifically breeding for certain colors. (IQHA has a registered champagne roan horse, the first documented of this color combo in any American breed! They also have black tobianos, black greys and a black roan [blue roan in Quarter Horses], etc.)

The head should be refined and reflect alert intelligence. Nostrils should be large and sensitive. The muzzle should be small with a firm mouth. Well developed jaws give the impression of strength. Quarabs may have a slightly concave, dished facial structure, hinting of the Arabian ancestry without being extreme; the amount of dish will vary with the type of individual. The head should demonstrate quality and beauty.

Eyes should be large and expressive; they are generally dark but may be blue (in double dilutes or pinto horses) or hazel (in champagne colored horses). They should appear bright and clear and be set well apart on the head.

Ears should be small, finely pointed and curved in at the tips. They should be set well apart and carried alertly.

Manes and forelocks should be full. The throatlatch should be clean and well defined, never thick, allowing for easy air flow when working. The neck should be smoothly joined to the shoulder and deepest at the point of attachment. The neck should be long and refined but not so fine as to appear weak. Thickness of the neck will vary somewhat with the type of individual. The neck should display a natural arch. Stallions tend to have a more fully developed crest than either mares or geldings.

The chest should exhibit good depth and width. Ribs should be well sprung and close. Heart girth should be wide and solid to allow for ample lung and heart space. The barrel should be large and round, yet trim in the flank. The overall body should be deep and full in the mature Quarab.

Withers should be well defined and of medium height, slightly higher than the point of the hip, and should extend well back. This allows them to hold a saddle well. The shoulder should have a deep slope, be well muscled and of good length.

The back should be compact, broad and deep in the girth. It should be close coupled, full and strong across the kidney.

Legs should be long and sound, made up of flat bones and solid, large joints. The forearm should be broad and well muscled; it should taper to the knee. The Quarab should possess a short to medium length cannon bone that is free of meatiness. Tendons should lie smoothly over the bone. They should appear wide and strong when viewed from the side. When viewed from the front, they should be straight, thin and set well apart. The hind legs should be placed so that the Quarab is able to turn on its hindquarters with the legs well under the body. Pasterns should be clean and strong, short to medium length and their angle should match that of the shoulder.

The hip should be quite muscled with a somewhat horizontally built pelvis. When standing with hind legs slightly apart (as in a halter class), the croup should appear nearly horizontal with little slope. However the croup should not be extremely flat as in many purebred Arabians. When moving or at work, the croup will appear more sloped, allowing for good engagement of the hindquarters. Overall the hip and hindquarters should show great strength. The hind leg should be muscled inside and out to provide good driving power.

The hocks shall be neither close together nor wider apart than the fetlocks when viewed directly on from the rear. They should be clean, wide, strong and straight.

Hooves should be medium sized, nearly round and open at the heel. They should be smooth and dense, never brittle. Hoof size should be adequate for the size and weight of the individual.

The tail should be set fairly high, with a natural arch and carried gaily, especially at the trot.

Movement will vary with the individual, depending on its type and breeding. However all Quarabs should move freely. They should work well off their strong hindquarters and move in a collected manner. While the amount of "action" will vary with breeding type, extreme "park" type action is to be avoided.

TYPES:

Body types may be customized, as there are definite preferences among breeders for the different types. As an example, western ranches tend to prefer the stock type with higher percentages of Quarter Horses and Paints. Endurance riders prefer the pleasure type, favoring the Arabian. But the majority prefers the 50/50 Straight type, which is a blending of the best of both worlds.

There is something for everyone with this diversity in types.

Straight or Foundation Type animals should be a good blend of the traits of both the Arabian and the Quarter Horse/Paint. They should not be overly refined but should exhibit a strong influence from both blood-lines. The Quarab Standard is that of the Straight Type Quarab.

Stock Type animals will demonstrate more of those traits commonly associated with the Quarter Horse or Paint but should retain an elegance and typiness from the Arabian.

Pleasure Type animals will more closely resemble their Arabian ancestry with more refinement, especially in the head. However, the body should still show a strong influence from the Quarter Horse or Paint blood.

Many owners are interested in registering their half Arabian/Quarter Horse or Paint blend horse because it goes beyond the Arabian Registry, which only records the Arabian half of the pedigree. As a registered Quarab, a horse's full heritage is recorded. This is a definite asset to those who appreciate the total breeding of their horse to be recognized.

As for the Quarab temperament, Tishyno's owner, Hans-Peter Marquardt comments, "I am happy to own this unusual horse with a cool mind and a big heart. Tishyno always tries to do his best. He is a perfect allrounder horse. He can do it all and is now a famous stallion."

Kristi Kraling agrees about Quarab temperament. She likes to talk about her Quarab gelding, who loves people and trail riding so much that he cries with anticipation whenever he sees or hears a horse trailer pulling up to the stable.

The Quarab combines the best of both the Arabian and Quarter Horse, and therefore is the best of both worlds.

The American Azteca

This breed inherits beauty, temperament, pride, agility, and spirit from the Andalusian and strength, heart, and speed from the Quarter Horse. The possibility of Paint coloring also brings added flashiness to an already spectacular horse.

The Azteca breed originated in 1972 when the Mexican charros (cowboys) began a quest to produce a horse that would represent Mexico as their contribution to the equine race. They required a horse with agility, quickness, and cow sense to work on their cattle ranch-

soon became the next destination for the breed. It was there that the breed took an interesting turn: it developed in a slightly different manner, yet its bloodlines continued to be based on the original Azteca of Mexico. Still a combination of Quarter Horse and Andalusian, the American Azteca also allowed "Quarter Horses with color" in its breeding. Since the Paint Horse is derived from the Quarter Horse, essentially it is a Quarter Horse, but with color, and could be used to produce American Aztecas. To qualify for breeding, the

es and yet retain the elegance needed for exhibitions, rodeos and parades. For this they chose the Andalusian to cross with their Quarter Horses and Criollo mares.

The result was astounding! The horse that emerged was gifted with athleticism, a willing attitude, stamina, grace, riveting beauty, harmony of form, outstanding disposition and a great talent to learn. Not only did it possess the ability to work on cattle ranches, but also the versatility for many other uses. The Azteca was born, and in the years following it acquired so much recognition, it earned the title as the National Horse of Mexico.

It became such a popular horse that inevitably it was noticed by other countries. America was among the first to appreciate the Azteca and

Quarter Horse or Paint Horse couldn't have more than 25% Thoroughbred blood.

Foundation Breeds: Quarter Horses and Paint Horses are breeds originating in the United States and are great athletes known for their use as pleasure horses, but are particularly renowned as working ranch horses. They are quick, maneuverable, and even tempered. They have a natural cow sense and often work cattle without much guidance from the rider. They are attractive and compactly built with large powerful hindquarters, strong shoulders, and short, muscular backs. They became popular and were named for their quick burst of speed at a quarter mile race.

The Andalusian is an ancient and rare breed. It is very strong with long sloping shoulders, natural collection, and extremely sturdy legs and hooves. Sought after for its quiet temperament and extreme intelligence, it is easily handled, yet has a reserve of energy when called upon.

Throughout history Andalusians were revered for their abilities as warhorses. These same skills were used in Spain and Portugal to work cattle and the notorious fighting bulls. Andalusians are still used in those capacities today, carrying their riders in the bullring with unimaginable grace and speed. The Andalusian is the working ranch horse of Spain and Portugal, just as the Quarter Horse is in America.

Today, 80% of all modern breeds trace back to Spain and Portugal's illustrious Andalusian, including the Quarter Horse and Paint Horse. So utilizing Andalusian blood would not be something new, but a reintroduction of a bloodline that is already present in the Quarter Horse.

Breed Standards: The American Azteca combines characteristics of both the old and new worlds of Andalusian and Quarter Horse, resulting in a noble, docile, agile, proud, and spectacular horse. It should have a good balance between the two breeds with qualities of both. The intention is to create a new type of horse, a new breed that exhibits the best of both ancestor breeds.

With some allowance for variations, the recommended characteristics of The American Azteca are as follows:

Size ranges from 14.2 to 16.0 hands. Both Quarter Horse and Paint Horse markings and colors are acceptable.

The head is of medium size with a straight, slightly convex or slightly concave profile. It has a broad forehead, expressive eyes and medium ears, which are mobile and well placed. The neck is well muscled, shapely, and slightly arched with a medium crest and broad at its base where it joins onto a long sloping shoulder.

The withers are broad and slightly muscled, yet defined. The haunches are strong and well muscled, leading to a medium to low-set tail. The legs are well muscled with dense bone, good joints and strong hooves. A long flowing mane and tail are often apparent.

Retained from the Andalusian is a free and mobile shoulder and hip, which allows the Azteca to be incredibly athletic and smooth to ride. The movement is naturally collected with a variance of knee action from high and brilliant to long and flowing.

The American Azteca responds exceptionally well to the different equine High School disciplines requiring suspended and elevated gaits. The qualities passed on from both parent breeds also make it a skillful working cow horse or western horse. It can and does excel at many events making it an extremely versatile horse. The breed is very easy to train and once taught, never forgets. This combination creates a horse anyone would be proud to own.

The American registry greatly respects the Mexican standards as Mexico was the original developer of this magnificent breed for its use there. Because it wants to credit Mexico for the creation of the Azteca and doesn't want to interfere with their national breed, the American registry doesn't call Aztecas outside of Mexico as Azteca, but rather The American Azteca. This signifies that they are the American version of the fabulous Azteca originally created in Mexico.

Although the Azteca began as a breed in Mexico, the culture and tastes in America called for a slightly different type of horse to fit its needs. The American Azteca Horse International Association (AAHIA) takes into consideration the demands, requirements, market, and abilities of breeders and owners of Aztecas in the United States and throughout the world.

Yet the AAHIA horses are modeled closely after their Mexican cousins in type. The organization still bases the breed on the combination of Quarter Horse and Andalusian blood and promotes a high quality horse. It doesn't allow more than 25% Thoroughbred blood in any Quarter Horse or Paint used to produce an Azteca. The AAHIA traces back 4 generations, not including the horse itself, for Thoroughbred blood. If those generations have more than 25% Thoroughbred blood, that horse is not allowed for breeding an Azteca. The registry does reserve the right to research further generations if more Thoroughbred blood is suspected.

Also, horses applying for registration have to be tested for HYPP (Hyperkalemic Periodic Paralysis Disease) if they have the Quarter Horse "Impressive" bloodline, and they won't qualify unless the test is negative. (Tests results may be: a.) positive, meaning the horse has it, b.) negative, meaning the horse doesn't have it, or c.) negative/positive, meaning the horse may or may not get it, but may be a carrier, in which case the horse won't qualify either.)

The goal of breeding American Aztecas is to arrive at a point where a new breed is created. To do this, several generations must be bred. The first generation of an American Azteca is to breed an Andalusian to a Quarter Horse or Paint Horse and that offspring is given the notation of an American Azteca D. When that American Azteca D is then crossed back to an Andalusian, this is an American Azteca B or second generation. If a D Azteca is bred to a Quarter Horse, that foal is a C offspring. When both a D and a B are bred, that is an American Azteca A. This generation then has been produced by breeding American Azteca to American Azteca and was not a cross of the two ancestor breeds… and a new breed has begun. The A is significant since the breed comes in A, B, C or D and the A is the 3rd generation or a result of breeding Azteca to Azteca.

There are other combinations that can be used to also arrive at an Azteca A, but this is the fastest and can take 12 to 15 years to achieve. It also gives a great blood percentage to the A, that being 5/8's Andalusian and 3/8's Quarter Horse. A breeding chart is available to show the different letters of an American Azteca and how to get them, or in other words, what crosses are allowed. The letters of A, B, C and D are not indicative of quality. There are quality horses at all levels and all are considered American Aztecas.

Noted American Azteca breeder, Rita Greslin-Ricard, states that it has taken her about 15 years of breeding American Aztecas for her to arrive with an Azteca A horse. At the moment, they are very rare, probably less than 15 Azteca A's of any sex in the U.S. and of those, less than 5 are Azteca A stallions. She counts it as a real privilege to be one of the first breeders to have bred and raised her Azteca A stallion, "Primero Viento", who is registered as Number One in the registry. Eventually he may be remembered as a significant contribution to the American Azteca breed, just as "Whimpy", Number One, is in the AQHA. Primero Viento

was produced from an Azteca B stallion and a D mare, so he is the coveted 5/8's Andalusian, 3/8's Quarter Horse percentage.

Rita has been an Andalusian breeder since 1980 and has tried many Andalusian crosses with other breeds. She states, "I can honestly say that I have yet to find an Andalusian cross that I don't like, but the American Azteca definitely stood out to me. They are just the perfect type of horse for any purpose. They are wonderful to be around, kind, sensitive, talented and willing are words that come to mind. They're EASY to own."

She believes the future of the breed is very, very promising. "They have everything that anyone could desire. The Quarter Horse started from crossing different breeds. It also started out small and blossomed into the huge influence it is today because of its talents. I believe the American Azteca has even MORE talents. I expect to see great things with this breed in the future."

Rita's American Azteca stallion, "Vaquero", who is the sire of Primero Viento, won National Champion Western Pleasure, Reserve National Champion Hunt Seat and Top 5 in Dressage Suitability at the Andalusian National Show. In classes where he competed against both Andalusians and partbred Andalusians, he won First Place in Doma Vaquera (Spanish reining which is a combination of reining and dressage) and First Place Trail Class. He also won several West Coast Championships and a Bronze medal in dressage at a USDF show.

"That tells just HOW versatile and talented this breed is.," she says. "You can do it all with ONE horse!"

So whether the need is for English or Western, graceful dancer or working ranch horse, jumping, dressage, cutting, reining, penning or just a great companion for trail riding, The American Azteca is a wonderful choice. They can do it all and do it well!

Credits

Following are the organizations, professional breeders, photographers and photo contributors who supplied the facts and photographs. Without their help, this book would not have been possible. To them go many heartfelt thanks.

Breed contributors:

Akhal-Teke:
The Akhal-Teke Association of America, Inc. MO
P.O. Box 1635
Rolla, MO 65402
akhalteke@fasterlink.com

Photographers and photo suppliers:

The Akhal-Teke Association
of America, Inc. MO
and: Linda Aaron
and: Cascade Gold Akhal-Tekes
www.cgakakhaltekes.com
and: Magic Valley Ranch
www.magicvalleyakhal-tekes.info

American Bashkir Curly:
Marni Malet, Bearpaw Ranch
and: American Bashkir Curly Registry
P.O. Box 151029
Ely, NV 89315
mmalet@libby.org

Greg Oakes
RR #5
Guelph, ONT. Canada N1H 6J2
www.curlyhorse.com

American Paint Horse:
American Paint Horse Association
P.O. Box 961023
Fort Worth, TX 76161-0023
askapha@apha.com
and: D. Phillip Sponenberg, DVM, Ph. D.
Director of Student Affairs
Professor, Pathology and Genetics
Virginia-Maryland Regional College
of Veterinary Medicine
Virginia Tech, Blacksburg, VA 24061

The American Paint Horse Association

American White Horse and American Creme Horse:

Carley Daugherty

and: The American White Horse Registry,
   or AWACHR
   90,000 Edwards Road
   Naper, NE 68755
   carleyd@juno.com

White Horse photo: Don and Jo Ann Anderson
Texas White Horse Ranch
18730 FM Rd. 15
Troup, TX 75789
txwhr@earthlink.net

Creme Horse photo: Carley Daugherty
and: Betsy Bailey and Bobby Catchings
14440 Old Humble Pipeline Road
Conroe, TX 77302

Appaloosa:

Appaloosa Horse Club
2720 W. Pullman Road
Moscow, ID 83843-4024
publicrelations@appaloosa.com

and: International Colored Appaloosa Assoc., Inc.
   P.O. Box 99
   Shipshewana, IN 46565
   icaa@aol.com

page 18: Appaloosa Horse Club

page 22: Alivia Dockery
Alivia's Appaloosas
Denver, CO
www.aappaloosa.com

AraAppaloosa:

Randy and Julie Berghammer, Sandy Hollow Farm
W4333 Cty. X
Markesan, WI 53946-7333
r4jsberg@yahoo.com

Claudia and Alexander Kaul
Most Colorful
www.buntearaber.de

Arabian:

Arabian Horse Association
10805 E. Bethany Dr.
Aurora, CO 80014
info@ArabianHorses.org

Jerry Sparagowski

Brindle:
J. Sharon Batteate
P.O. Box 8535
Stockton, CA 95208
jsbatteate@aol.com
and: VetGen
    3728 Plaza Dr., Suite 1
    Ann Arbor, MI 48108

J. Sharon Batteate

Buckskin:
International Buckskin Horse Association, Inc.
P.O. Box 268
Shelby, IN 46377
ibha@netnitco.net

page 35: Jill Johnston
J & J Paints and Quarter Horses
2525 Lease Drive
Dodgeville, WI 53533
www.angelfire.com/wi/painthorses

Grulla photo: Bryant Ranch
page 36 36401 S. 570 Rd.
Jay, OK 74346
info@bryant-ranch.com

Half Arabian and Anglo-Arabian:
Arabian Horse Association
10805 E. Bethany Dr.
Aurora, CO 80014
info@ArabianHorses.org

Half Arabian photo: Jan Wolf, Providence Farm
page 40 7024 Esker Rd.
Custer, WI 54423
http://webpages.charter.net/providencefarm

Anglo-Arabian photo: Peggy Ingles, Starstruck Farms
page 42 20 Lambourne Rd., #221
Towson, MD 21204
peggy@starstruckfarms.com

Morab:
International Morab Breeders' Association
and International Morab Registry ™
732 S. Miller Court
Decatur, IL 62521
Briscoe_j@sbcglobal.net

Rick Kroll
Box 982
Cochrane, Alberta
T4C 1B1
okkroll@telus.net
and: Prue Critchley
Box 487, Hamiota
Manitoba, R0M 0T0 Canada
www.bartongate.com

Morgan:
The American Morgan Horse Association, Inc.
122 Bostwick Rd.
Shelburne, VT 05482
info@morganhorse.com

The American Morgan Horse Assoc., Inc.

Moriesian:
Moriesian Horse Registry
1001 N. Russell Rd.
Snohomish, WA 98290
Moriesian@aol.com

Pauline Morin
HC01 Box 283
Pelkie, MI 49958
and: Ellie Neerdaels
Twin Artesian Stables
Green Bay, WI.
www.moriesian.com

Palomino:
The Palomino Horse Association, Inc.
Rt. 1, Box 125
Nelson, MO 65347

The Palomino Horse Association, Inc.
www.palominohorseassoc.com

Pintabian:
Pintabian Horse Registry, Inc.
P.O. Box 360
Karlstad, MN 56732-0360
218-436-SPOT

Jodi Blesener
13332 Friesland Road
Hinckley, MN 55037
www.grindstonepintabians.com

Pinto Horse:
Pinto Horse Association of America, Inc.
7330 NW 23rd Street
Bethany, OK 73008

Stock Type photo: McLochlin Show Horses
page 63 Harris Paints
www.harrispainthorses.com

Hunter Type photo: Abby Duncanson
page 63 Farmington, MN.
www.pinto.org/users/minnesota

Pleasure Type photo: Carrie F. Ferlita
page 64 Trimbelle, WI.
651-245-8576
and: Sandra Dungan Anderson
Sanbara Farm

Saddle Type photo: Laurel Reeves and Linda L. Wolff
page 65 Cambridge, MN 55008
www.paintedoaksfarm.com

Quarab:
International Quarab Horse Association
P.O. Box 263
Hopkins, MI 49328-0263
QuarabRegistry@aol.com

Straight Type photo: MQ horses
Kati and Hans-Peter Marquardt
Brettener Str. 68
75031 Eppingen/Germany
+49-170-31 33 205
www.mqhorses.de

Pleasure Type photo: Claudia and Alexander Kaul
Most Colorful
www.buntearaber.de

The American Azteca:
The American Azteca Horse International Association
P.O. Box 1577
Rapid City, SD 57709
dwinds@iw.net

Diana Allison, Full Moon Photography
www.equuphoto.com
and: Dakota Winds Andalusians &
American Aztecas
Rita Greslin-Ricard
Sturgis, SD
www.dakotawindsandalusians.com

Cover photo: Cheyenne River Ranch
cc2cda@gwtc.net
and: Diana Allison, Full Moon Photography
www.equuphoto.com

Inside front cover: Sherry Lytle, Egyptian Image Arabians
Rapid City, SD 57703
and: Diana Allison, Full Moon Photography
www.equuphoto.com

page VIII: Ellie Neerdaels
Twin Artesian Stables
Green Bay, WI.
www.moriesian.com

page IX: Don and Jo Ann Anderson
Texas White Horse Ranch
18730 FM Rd. 15
Troup, TX 75789
txwhr@earthlink.net

page XI: Susan Lord
and: Lisa Banks
203 W. Trent St.
Trenton, NC 28585
lottaspots@hotmail.com

Look for upcoming books in this series:

Horses of the Range: All American Naturals

Fleet and hardy breeds from the panorama of wild and open ranges.

Ponies and Small Horses: Little Pals with Big Hearts

Who can describe the relevance of a child's winsome pony companion or the hard working faithfulness of an enduring little horse?

The Gaited Greats: Free and Easy Movers

From the expanse of green plantations to icy mountain tops, these superlative riding horses established themselves as the epitome of traveling pleasure.

Powerhouse Horses: The Movers and Shakers

From the brawn of spectacular drafts to the muscle and ability of others, a job is never too big.

The Classics: Equines of European Antiquity

Renowned athletic steeds from ancient Europe flourished through the ages.

HALLELUJAH PUBLICATIONS

www.equestrian-horses.com • Direct 715-273-2135 • 800-533-1635 ext. 135